Synthetic Friends

Hendrik Kempt

Synthetic Friends

A Philosophy of Human-Machine Friendship

Hendrik Kempt
Applied Ethics Group
RWTH Aachen University
Aachen, Germany

ISBN 978-3-031-13630-6   ISBN 978-3-031-13631-3   (eBook)
https://doi.org/10.1007/978-3-031-13631-3

Cover illustration: British Modern Photography

This Palgrave Macmillan imprint is published by the registered company Springer Nature Switzerland AG
The registered company address is: Gewerbestrasse 11, 6330 Cham, Switzerland

For my friends

ACKNOWLEDGMENTS

I would like to extend my gratitude for helping me begin, write, and finish this book to a number of people. I began this project at the height of the second wave of the Covid-19 pandemic in Germany, in which social contacts have been severely limited. I spent most of my time at home, like most people in Germany, Europe, and the world. This has, without doubt, shaped the views that I am defending in this investigation. What began as skepticism about the often emotionally laden rejections of the idea of human–machine friendships has bloomed into a rather extended defense of the possibility of such friendships.

During this time, I had the opportunity to present my thoughts on several occasions to my colleagues, whose feedback made clear that a defense of certain premises is an apparent necessity for any such project. Saskia K. Nagel, Jan-Christoph Heilinger, Niël Conradie, Peter Königs, Nils Freyer, W. Jared Parmer, and Chaewon Yun helped me shape the structure for such an extensive investigation, to keep an eye on both the fundamental issues as well as the more fine-grained detail. For this, I owe them my thanks. However, I have to direct a special thanks to Saskia K. Nagel for her understanding of the time and attention necessary for this project to be finished.

I would also like to thank my students of the course I had the pleasure of teaching in the Summer term of 2022 on human–machine relationships at RWTH Aachen. In this course, we had fruitful discussions about all sorts of tricky questions about the sociality of machines and our ability to enter relationships with them. Also, I apologize for making you read one of my papers on human–machine friendships to help me relate my thoughts to a broader audience. In this context, I would like to thank David Gunkel for taking the time to defend his approach to the moral patiency of machines in my course.

Special thanks belong to artist Zeno Gries, whose artistic engagement with AI in the form of a project on natural language processing and my inclusion in this project motivated me to reflect upon my assumptions of what AI will do in the world, how it can be used for pushing our understanding on different levels, and how I can play a role in it outside academia.

I would also like to thank Rachael Ballard and Uma Vinesh for seeing the process of this whole project through with patience and advice. Without them, this book would have never been finished. Further, thanks belong to the anonymous reviewers of the initial proposal, whose helpful and constructive criticism put the project on a clear path toward completion, and John Danaher for the endorsement of the finished book.

My family owed thanks for tolerating my life choices for the last decades, even though they are rarely comprehensible to them. Lastly, thanks go out to my friends, both digital and analog, for teaching me about being a better friend, for being there when needed, for the encouraging words, the helping conversations, the wine, the time, and the jokes and stories. Thomas, Lucas, Melis, Kerstin, Linus, Isabella, Alon, Ahmed, Max, Edu, Taner, Danny, and Ardalan, and of course always John. You are the very definition of a chosen family. Thank you.

CONTENTS

CONTENTS

CHAPTER 1

Introduction

Friendships are such a common phenomenon in human life that trying to explain them philosophically can often miss the very point, lived experience, and meaning of those relationships for those in them. Similar to an analysis of humor, which usually sacrifices the fun and amusement one can have with jokes, a philosophy of friendships is risking to sacrifice the unspoken easiness with which we often have relationships for a however decontextualized or artificial understanding of what friendships mean.

However, there are some truths, or at least intuitions, to friendship that might motivate a philosophical investigation. For example, the statement that "some friendships are better than others". This rather intuitively correct assessment provokes a whole range of other questions that we should then have answers to. What does it mean to be in a "better" friendship? Are we not simply in as good a friendship as we ourselves are good? Who decides which friendship is better than another? Should we try to be in better friendships, even if we disagree with the conditions of what makes them better?

Philosophy has given answers to these questions, though we should be cautious to not kill the joke by explaining it. The most consequential theory to this day is Aristotle's approach to friendship. It has been discussed and reproduced in other classic works of Western philosophy like Cicero's "De amicitia" (Cicero, 2014), and has experienced a shift

© The Author(s), under exclusive license to Springer Nature
Switzerland AG 2022
H. Kempt, *Synthetic Friends*,
https://doi.org/10.1007/978-3-031-13631-3_1

within the Christian tradition by the writings of Thomas Aquinas (in his "Summa Theologiae") toward a more generalized, less particular idea of relationships. In prioritizing that our love should first be directed toward God, Aquinas stresses the relevance of agape in these non-romantic relationships over philia, the love between humans.

Whether one agrees with Alexander Nehamas, who concludes that our contemporary understanding of Aristotle's original idea is thus flawed due to the bait-n-switch of exchanging the partial love of philia for the impartial love of agape, or whether we prefer to read these ideas from a generalizable account—the fact that Aristotle's theory is dominating theory is undeniable.

The development of artificial intelligence and robotics over the last 15 or so years and the development of the internet toward a thoroughly social medium have provoked serious philosophical interest in exploring the future of our relationships. And knowing Aristotle's current influence on the philosophy of friendship, it is not surprising that it has been applied to our latest insights into the ability to build machines with which we can build relationships. The discourse began with a struggle to assess whether the internet on its own can be a medium with which we can conduct friendships of this virtuous kind. We add to the already rich discourse in favor of almost any kind of friendship being purely mediated through technology by meditation on hermits, especially digital hermits. At this point in our technological development, not only can we live a rich, fulfilling social life purely online. It also seems that serious discussions about human–machine friendship are to be had.

From the subject of "friendship through technology" to "friendship with technology", philosophical attention has now been directed at the question of whether machines will ever be plausibly our friends. And still, Aristotle's account is the ever-present premise from which most philosophers operate.

This, however, can have somewhat problematic consequences, which can be illustrated in the line of argument pursued in John Danaher's account of robot friendship (Danaher, 2019) and Sven Nyholm's criticism of it (Nyholm, 2020). Danaher presents a theory on how to interpret Aristotle for human–machine friendships (we will get to a more elaborate analysis in Chapter 5), with features of what makes a virtue friendship, and answers whether machines will be able to fulfill these (Danaher thinks they can). Nyholm disagrees and argues why some of the features given by Danaher must mean that machines cannot become virtue friends, and asks a bigger question, which is whether we should even want these machines to exist. What both seem to take for granted is that we might want to start much lower when asking whether human–machine friendships are possible. Nyholm even grants that we can benefit from machines in the same way that we can profit from lesser friends, and just stops short of admitting that, in some meaning, human–machine friendships are possible.

This kind of discourse is the consequence of Aristotle's account. He even says as much, as he also only recommends concentrating on virtue friendships. Especially in the question of whether machines will be able to join this already small circle, the discussion seems to condense into some debate about whether this or that condition can be fulfilled by machines. Yet, admittedly, if we reject the quest for virtue machine friends, we are left to answer what else there could be.

One of the main concerns about human–machine friendships, because of their striking intuitive plausibility, economic probability, and relative simplicity, is that the creation of social machines will mean the creation of insincere pleasure machines. We get much mileage out of the assumption that the design of machines to fit our purposes must lead to the creation of machines that merely please us. The challenge of rejecting this proposal involves an elaborate idea of what a virtuous machine friend would do (in contrast to a mere pleaser) and how we can avoid the pitfalls of slipping back to the more straightforward solution of a pleasure machine.

In this investigation, we pursue a double strategy. On the one side, we acknowledge that a debate on whether this or that property is necessary or sufficient for a virtue friendship is misguided. On the other hand, we still should find features of a machine that avoid the worst outcomes, maybe to the degree in which John Sullins suggested letting the market create its own machines (Sullins, 2008).

This double strategy leads to proposing a theory of synthetic friends that are defined mostly by their negative features. We take for granted that social machines will exhibit some of the relational properties deemed relevant by the moral community (cf. Gunkel's, Darling's, and Coeckelbergh's work on the matter), but settled on negative necessary conditions afterward. Non-deceiving, non-exclusive, non-coercive, and non-criminal/violent social machines may be all it takes in this sense to build any kind of friendship with them. However, once we accept that positive suggestions are in danger of pushing the creation of pleasure machines, we might be able to make some constructive suggestions that endorse the

difference between machines to humans and thus create the possibility for
the field of human–machine friendships being somewhat detached from
the debate on virtue friends.

Thus, we should utilize the unique abilities of machines (similar to us
using the unique abilities of pets when relating to them) that can have
beneficial effects on our lives and relationships without deskilling us or
manipulating our social expectations into unhealthy standards for other
people. We settle on calling these machines "synthetic friends".

As an admission that friendships do exist within a larger social context,
we ought to explore the ability of machines to be socially integrated. This
appears to be the biggest concern for human–machine friendships: if we
can add synthetic friends to our friend circles, the worries about deskilling,
atrophying, or corroding social skills seem inadequate.

One may claim that the structure of this book, reflecting both the
negative approach to friendship as well as the challenges that bear from
attempts at applying them to social technology, is front-heavy. However,
as we see it, it is reflective of the work needed to avoid the severe
debates one witnesses on other areas of advanced human–machine interactions and their normative consideration: the robot rights debate and the
controversy about sex robots both suffer from a lack of such initial justification (or, more likely, are simply misunderstood). Research is being
attacked on the mere grounds that it exists, as thinking about the topic
of robot rights, e.g., strikes others as an offensive waste of resources. The
idea here is to make it as difficult as possible for people to misunderstand
what our debate is about, but also acknowledge that there is indeed a
burden of proof to be lifted for such an investigation.

Thus, we must begin with a lengthy justification of this book. While
some concerns should be addressed that do not at all affect the contents of
a human–machine relationship but the conditions under which those can
form, some others do not. The ones we recognize as worth dealing with
are both philosophical as well as empirical matters. From purely ontological concerns about the aptness of the category of "friendship" for
machines to the economic contexts in which machines are being built to
the political landscape of legal regulation—these will affect the potential and content of human–machine friendships. However, the lack of
technological sophistication is not such a reason.

Much of what will have been said in these chapters is controversial.
To anticipate some of the arguments brought against these claims, I will
turn in the last chapter to some of the challenges of our proposal. Several

charges will prove to be troubling: over-inclusivity on a conceptual level, the ability to be sincere with machines, some existential and aesthetic concerns, some ethical concerns about the justifiability of such machines in the first place, and apparently absurd implications of machine–machine friendships.

First, the over-inclusivity might water down the concept of friendship to a degree that it is hollow, though we may confirm that there are indeed still some positive features to be had and discussed that suggest sufficient specificity. Second, allowing machines to be friends of this kind might cause some authenticity/sincerity issues we might rely on in thinking about friendships. We show here that sincerity can be understood pragmatically to be "unwithholding", and thus include machines. Third, the issue of non-aging machines and their lack of a concept of personal finitude is supposed to make them less relatable. After pointing out that we rarely relate to each other based on our shared existential finitude, this argument seems somewhat overly intellectualized. Fourth, the ethical argument against machine friendships, discussed by Nyholm, contends that the capacities of these machines, while useful, will have a net negative impact and thus should not be constructed in the first place. We hope to cover this argument by presenting a much smaller account of what machines are supposed to be able to do. Fifth, the seemingly absurd consequence of machine–machine friendship can be weakened by referring to the lower-level concepts of friendship we discussed before.

In refuting these arguments, it stands to reason that the account presented here is viable to the degree that no specific philosophical argument emerges that would not affect all other accounts of any human–machine friendship, which we discuss at the very beginning of the book.

The goal of this account is to build a case for human–machine friendship that can integrate machines into people's lives as friends. We hopefully achieve that the problem of strictly rejecting human–machine relationships and denying people's lived reality will cause harm. We should be prepared to offer explanations as to what kind of relationships people are about to enter.

From there on out it is left to any single person whether they would ever consider these relationships for themselves. And even while the answer may remain a firm "no" for most, the secondary, normative goal of this investigation is to compel people to accept others entering human–machine friendships. They do this for their personal reasons, and what this investigation shows is that there can be good reasons for seeking such connections. We fare better in trying to understand these reasons.

REFERENCES

Cicero. (2014). Laelius. Oder: De Amicitia. Reclam.

Danaher, J. (2019). The philosophical case for human-machine friendship. *Journal of Posthuman Studies, 3*(1).

Nyholm, S. (2020). *Humans and robots: Ethics, agency, and anthropomorphism.* Rowman and Littlefield.

Sullins, J. (2008). Friends by design: A design philosophy for personal robotics technology. In *Philosophy and design* (pp. 143–157). Springer.

CHAPTER 2

Why Even

A book concerning itself with the question of whether we can become friends with technology should own the burden of proof to show that not only is asking this question appropriate but also that there is a chance to come to a productive, positive answer. It appears that rejecting the question as well as any potential positive answer comes much easier and quicker to most people. The very thought of becoming "friends" with machines in any sense of the word may be Science Fiction at best, and dangerous at worst.

Though this book is not the first one to combine the issues of friendship and robotics (or, as we will see later on, generally more sociable machines), it still appears that any work on this issue has to carry the burden of proof of intelligibility, utility, and seriousness. While eventually, the amount of good scholarly work likely will carry its own weight and thus establish a branch of thought on its own, at this stage of philosophical inquiry, we may still find it necessary even to motivate the very research object that is human–machine friendships. That is because inquiries in this field, and similar areas of human–machine relationships, are often met with a split audience. One side may accept the premises of the inquiry and engage in arguments; the other side may reject those premises as utterly misguided, pointless, or even offensive as it ignores the fundamental injustices of technology production.

© The Author(s), under exclusive license to Springer Nature Switzerland AG 2022
H. Kempt, *Synthetic Friends*,
https://doi.org/10.1007/978-3-031-13631-3_2

To engage both sides equally, we may then first want to work a bit on the common ground that such an inquiry starts from. Thus, a defense of "why even talk about this?" is needed to not lose many philosophers of AI, computer scientists, and other scholars from the start. This defense of human–machine friendships as a branch of thought still may turn out to be false or unconvincing; however, we may still learn from such failure why it failed—the opposing arguments are plenty but also not mutually inclusive. This means that even knowing which argument against human–machine friendships as a philosophical program are the decisive ones teaches us something. If, however, such an attempted defense of the very branch is successful, and others might have already shown that such a defense can justify this branch so it can carry itself, then any positive proposal stands on firmer grounds.

As we intend to provide an approach to human–machine friendship that is ambitious in its scope and must offend some people (as have several stages of the draft of some of these chapters), we take the distribution of the burden of proof to be fair and thus, this project is in need of extended justification. For this, I see three challenges that extensively inquiries into human–machine relationships ought to overcome or at least address, which can be generally called the "conceptual issue", the "ethical issue", and the "empirical-political concern":

1. the conceptual issue aims at the *concepts of friendship and sociality*; they are thought to be intimately connected with each other and exclusively within the human circle; thus, the question itself is fundamentally misguided,
2. the ethical issue means that approaches to human–machine friendship concern themselves with a technical solution to such fundamentally social ontology problems (from 1), thus falling into *technosolutionist* territory, and
3. the empirical-political concern that the world is not, and most likely will not be, of the kind needed for any of these questions to become relevant practical issues in the foreseeable future.

2.1   THREE CHALLENGES

These challenges merely affect any kind of positive theory for human–machine friendships. These concerns may return in some modified form in actual theories for human–machine friendships. Still, without providing convincing answers to these three challenges, no theory can demand to be taken seriously philosophically. We will discuss these three conditions of "why even" in more detail here.

2.1.1   First Challenge: Social Ontology for Whom

The first challenge is one of the social ontology intuitions. The issue taken with theories of friendship being expanded to artificial entities consists in this being a substantial expansion of not only the social circle but the moral one, too. One cannot, in any more theory-based sense of the term friendship (which is what most scholars aim to get at), merely claim that humans and machines can be friends without having to either comprehensively redefine the term friendship or to accept implausible consequences. In order to ask whether machines and humans can be friends, one has to, before that, ask "in what sense of friendship"? Yet, questioning and expanding the concept of friendship often looks like a misunderstanding rather than a fundamental redefinition. Instead of appealing to the need to redefine the concept and thus our social ontology, the diagnosis is that we are simply mistaken in our social ontology if we ask questions of friendships to machines.

One can point toward the fact that any philosophical investigation thus far presumes that friendship, embedded in social contexts, requires human beings. Thus, any given concept of friendship is based on assumptions of human friendships, rendering the question of whether humans and machines can be friends in this sense meaningless. The German term "Zwischenmenschlichkeit" can be understood to reflect this categorical distinction: it denotes the "interhumanness" of relationships, the shared humanity as a connecting point for social relationships. A machine, lacking humanity, cannot create or participate in "interhumanness" and thus cannot take part in friendships either (cf. for a discussion of this concept Kempt, 2021).

As strong as this charge is, if this was all there is to say about human–machine friendships, we would be finished already. Such a straightforward rejection of the research question as a categorical mistake appears somewhat inappropriate considering that especially social ontology is one of the evolutions and marked by growing inclusivity (especially in light of who Aristotle thought would be able to participate in virtue friendships). This is not to say that we may conclude that human–machine friendships are indeed best understood as conceptually something completely different, if not even undeserving of the term "friendship" due to those human-centered implications. However, methodologically it appears prudent to be careful with these strong judgments about the contents and extent of the social circle. Several authors have made this point strongly in a debate that faces related conceptual charges and divides: robot rights. The history and pseudo-rational method of othering groups of people and thus excluding them from any kind of moral consideration have set the precedent that social ontology is entangled with the power structures of a given time period. Those entanglements may even be implicit to us at this stage and will only reveal themselves upon reflecting years and decades from now. As we will get into the history of the social ontology of friendships, comparisons to the rights debate will reemerge as well. However, for now, we may point out that the concept of friendship has experienced an extension. Aristotle was rather explicit about the possibility that women and enslaved persons could achieve the levels of virtue required to reach the levels of friendship he considers of the "highest quality" (Aristotle, 2000), while nobody would seriously argue for that today.

Lastly, independent of what the inquiry will bring, the inquiry itself can be considered worthwhile from the potential lessons and insights into the matter of social ontology. Suppose we only find that the social ontology of friendship requires a rejection of the potential of any meaningful human–machine friendship. In that case, this presents an insight into the requirements of how far social ontology can be bent before it breaks. It is, however, decidedly not a pointless endeavor, nor should it be considered a categorical mistake. As it stands now, humans are more than willing to extend their moral and social considerations to non-human entities. From the outset, it does not seem unreasonable to think that the question of whether we eventually have strong movements to include machines in our friendship considerations, is indeed an open one.

2.1.2   Second Challenge: Technosolutionism

Next to the conceptual concern that friendship is too narrow a social ontological concept to accommodate a theory of human–machine friendships, we ought to concern ourselves with the charge of promoting technosolutionism. A solution is representative of a technosolutionist mindset if a technological answer is given, preferred, or considered ideal in the face of a non-technological problem. The "technosolutionist mindset" (Morozov, 2013), then, suggests that most problems we face today are merely not solved yet because of the lack of appropriate technology, including patently social issues like inequality or poverty.

This does not mean that technology can never be the solution to any non-technological problem, as the introduction of certain technologies had windfall effects in non-technological areas (think, for example, the automation of laundry machines, freeing up the time and energy at the time predominantly women workers to invest in other tasks). However, the concern with technosolutionist approaches is that we should not conceptualize social, cultural, or psychological problems or conflicts as solvable by specific technological inventions or general technological progress. Thus, in this view it is sufficient to find the right technology (or wait until the right technology catches on) to solve some of the most pressing issues, without having to involve ethical, social, or other non-technological approaches.

Further, it is often claimed that especially technologically mediated problems can be resolved by yet another technological invention. One of the main examples of such technosolutionism in current discourse is the right approach to climate change recommending "ecotechnology". No doubt, technological innovations will play a huge role in ameliorating the effects of climate change. However, many discourses about limiting the effects of human activities contributing to climate change consist of behavioral technologies, improvements in the efficiency of processes, replacement technologies in resource consumption, etc. What is missing is the social–political demand and insight that only a fundamental reorganization of many people's lifestyles will achieve the changes in carbon emission output that would limit the effects of climate change. However, such a reorganization is a political, collective choice based on deliberative insights, rather than yet another resource-saving technology that improves some production. Considering that all large-scale concerns are, effectively, of political nature, offering "apolitical" technical solutions may be correctly identified as technosolutionist.

Another area that is supposedly technosolutionist is the one of human–machine relationships. The argument goes as follows: An investigation into human–machine friendships should serve a purpose. Friendships serve a social purpose. If one is advocating for human–machine friendships, then one is advocating for a technological solution serving a social purpose. Thus, investigations into human–machine friendships by virtue of being an investigation alone could be construed to engage in technosolutionism. Any purpose machines could serve that resembles friendship could, and as some argue, should be taken care of by humans. The standard socio-psychological answers given to defend proposals are often of a pathologizing pattern: these friendships are intended for people with dementia, neuroatypicals, and children. For these use-cases in which human–machine relationships are helping the person to overcome some burden in their life, a machine may represent a medical solution more than a merely technological one. However, merely accommodating people who would just like to be friends without having to do the work for it appears to solve a social problem with a technological solution. Arguably, we ought to be more attentive to our fellow citizens, to be aware of potentially lonely, excluded, unhappy, or otherwise socially disconnected people. Inventing machines that "do the social work for us" strikes as an inappropriate "shortcut" that tells us more about our moral fabric rather than the feasibility of a technical solution.

This argument and charge can be rejected on two grounds. First, determining where a "purely" technical solution is appropriate and where it is technosolutionism is in itself a moral judgment. Such judgment ought to be made carefully and transparent in regards to a potential cultural or moral bias. Case in point, Jennifer Robertson's investigation into the role and status of robots in Japanese culture, and in Japanese families in particular, puts the force of the universalizability of such moral judgment into question (Robertson, 2018). While a culture with strong expectations and sensibilities about the closeness and interconnectedness of families, neighbors, etc. may judge that human–machine friendships introduce a technical solution where the culture calls for closer human bonds, it appears that other cultures would allow introducing machines into the social circle to take over some social tasks as well. However, are those cultures that are open to solving a social issue with technology technosolutionist in a bad sense? Certainly not.

There may be some universalizable limits that ought to be explored further, as some ethical standards of what we owe to each other may be violated when retreating from social situations and replacing them with technology. However, as this is, again, an open question and not one that allows a flat-out rejection of the research question, a technosolutionist charge against an investigation into human–machine friendships is not appropriate.

Second, technosolutionism still can lead to good results. Criticism against technosolutionism is often aimed at the intended technical solutions to a non-technical problem, claiming that this is a problematic misunderstanding of the role technology should play. Often, this claims is associated the misunderstanding of politics or collective responsibility to a problem. However, if we do attempt to incorporate technological solutions into our arsenal of possible solutions, and prioritize the pursuit of such technological solutions over others, it can on occasion yield positive results: either on its own or for enabling other solutions in combination with technological help. Take, again, the example of climate change. While a standard technosolutionist solution suggests that green growth eventually creates technologies that save on energy to the degree that is necessary to stop human effects on climate, they ignore the possibility that other solutions should be considered at the same time, or in combination with such technological approaches.

The problem here, often enough, is that technological solutions are seen as the only viable ones, mainly because other options—such as political or social accomodations—are more challenging to define, to reason for, are more uncomfortable, and less precise. Technological solutions, or rather their promises, suggest a certain time frame, allow for somewhat precise price estimation, and all that without having to touch many people's lives in a negative way. The allure to avoid having to demand substantial structural change by instead improving the given structure through an invention is rather obvious. But an interlocked approach between social change, facilitated and enabled through technological progress, might be more promising. For that, we need technological progress, some view of problems as primarily technological ones, in lock with the awareness that social solutions may be required to contribute to solving a large human issue.

For an investigation into human–machine friendships, thus, we should not focus on the question of application too much. This is tempting, as there are some already discussed cases in which cases the use of machine friends is appropriate. Mostly in the context of seniors and chronically lonely people, we find an intuitive appeal (and, at the same time, the very charge of technosolutionism emerging) for the utility of such machine friends. The main concern that once we need friends this badly, we should understand this issue as a social issue may be compelling enough to avoid a discussion about machines altogether.

2.1.3   Third Challenge: Contexts, Economic, Political, Social

The third major challenge to any kind of investigation into human–machine friendships consists in the necessity to consider the contexts in which social technology is produced. We can call this the "empirical-political concern" or the contextual challenge. The argument goes roughly like this: Many investigations into human–machine relationships, such as inquiries into moral standing, rights, etc., analyze the relational conditions and ethical demands of how we should treat certain robotic or otherwise individuated machines. These analyses, however, merely concern themselves with the content of our relationship with such technology, and rarely with the conditions under which such technology is produced. These production conditions, the aims, market incentives, power distributions, and others influence not only what kind of product is being made but also how it is made accessible to humans (see also Crawford, 2021 for an exemplary analysis of one of those devices). If we do not want to take any kind of human–machine interaction at face value, i.e., as a starting point for a relational analysis, we ought to consider how these machines came about in the first place.

Thus, an investigation into the possibility of being friends with these machines that do not account for the fact that humans create machines within the constraints and conditions of a specific political context and specific economical circumstance is missing a key element in the analysis. Missing this key element then renders these attempts to analyze human–machine relationships naive at best or ignorant at worst. Either way, many people forming this kind of argument claim that, if we were to consider these contexts and conditions, any analysis of human–machine friends would end up trivially rejectable. According to this presupposition, the often overlooked starting point of analysis would determine the result of the analysis. Thus, analyses, which do not use this starting point as necessary for any insight into human–machine friendships, are necessarily false or pointless. Thus, authors using this kind of argument fault their opponents for taking a decontextualized, and thus inappropriately idealized, view on these issues. In short, one cannot discuss social machines as if they are naturally encountered entities.

We call this challenge the "contextual challenge" since the debate focuses on the question of whether the context under which these machines are being constructed is predetermining the scope of any theory.

We ought to acknowledge that the context in which these technologies are produced, without doubt, plays a role in the scope and depth with which we can and should relate to these technologies. Some simple insights should lead quickly to this conclusion: relating to an artifact implies relating to the intentionality of the artifact, and thus the background with which it was constructed. A machine intended to trick you into relating to it in a way that makes you exploitable limits your ability to relate to that machine in any philosophically interesting way. If we were only able to build machines that try to exploit our personal data to target ads and thus improve the profit margins of the manufacturer, then any theory about human–machine relationships is not only technosolutionist but dangerously misleading about the potential of relating to machines. Clearly, we are capable of creating machines that are not that. Rather easily, actually. However, the contextual challenge reminds us of the very many different influences the construction of machines implies: technological constraints and capacities (such as reliable responses, maintenance, error-proneness, accessibility), economic demands and incentives (business models, profit margins, and stakeholder interests, market demand and competition, price), political and legal restrictions and requirements (data protection laws, youth protection, anti-discrimination laws, liability regulations, etc.), social and psychological necessities (uncanny valleys, age differences, personal propensities, pathologies, etc.), cultural contingencies and aesthetics preferences (different expectations about the required sociality of machines, aesthetic choices in design), ethical demands, and other conditions will lead to machines with a thick background of all sorts of normative relevance that may limit the potential for human–machine friendships.

However, considering all these reasons does not mean that an inquiry into human–machine friendships is pointless or not worth doing. These conditions should inform the inquiry into those relationships. We do have in most of these conditions not only a certain level of ambiguity and constant change but we also have a certain level of influence.

That the construction of social machines currently mostly lies in hands of private, profit-oriented companies who do not have an interest in creating machines that only function as friends to people gives reason to believe that the scope of human–machine friendships ought to be limited. However, this fact is neither necessary nor especially stable. We can imagine, and advocate for, publicly funded research with strong regulations into the ability to create social machines that mostly serve the public; we can further imagine standards of data processing and storage being proposed on those machines (as has been done already) in the political sphere strongly regulating the privacy of human users and interactants; we can imagine (or rather, point toward) a diverse range of cultural sensibilities that are more open toward incorporating machine entities into their homes; we can imagine the technology to mature to a degree where well-adjusted adults find use and liking in interacting with a machine on a friendly basis; we can imagine complex machine personalities that are challenging yet rewarding to interact with; we can imagine moral conventions that accept increased human–machine entanglement; we can, ultimately, imagine that human beings want to claim that they are in friendships with machines without having to decontextualize all these previously mentioned conditions. The fact that we can imagine the current conditions of social machines to change in such a way decreases the force of the "contextual challenge".

These conditions, however, must be a reflective dimension of a productive proposal on how to be friends with machines. As we have some however small influence over the way our research, industry, market, policy, laws, society, culture, and morals react to these machines, we ought to put any proposal on a strong empirical footing. Thus, exploring these conditions in more detail and providing suggestions on how to answer and incorporate these conditions will be the core of Chapter 3.

2.2   What Is Not a Challenge

Lastly, before turning to these empirical conditions in more detail, we should discuss one major point that is often brought forward that I do not consider a strong argument against human–machine relationship investigations: the charge that the technology is not "there yet" and that we do not know when or even if we ever get to the levels of sophistication such an investigation must presuppose in order to map normative constraints or social possibilities. The argument here relies on the ultimately primitive nature of the technology, especially the capacities of the AI, that partially motivates the argument mentioned before. Thus, taking human–machine relationships philosophically seriously must be either fully speculative or simply a misunderstanding of the available technology. Both options, in this view, do not provide the ground necessary for a philosophical investigation into human–machine relationships.

The main reason why I do not think that this concern, in contrast to the three ones mentioned above, is as big a challenge to any such investigation is the following: the actual technology available at this stage is not a relevant qualifier for the relevance of a philosophical investigation. It is true that many of the technological gadgets we are surrounded by thus far are not very convincing as social agents to a majority of human agents. It is also true that the progress of creating artificial conversational agents is rather unpredictable, with different, bigger companies working on large language models without publishing their results based on peer review. Many of the promises machine-learning made a decade ago did not hold, from the ubiquity and reliability of self-driving cars to the sophistication of personal assistants and the announcement of our relationships being transformed into human–machine ones.

Then, how is this reality-check of AI not conclusive to the futility of philosophical investigations into the prospect of the matter? Philosophy of technology is always, to a degree, a speculative affair, and it must be to be productive. Especially the ethics of technology ought to provide guarding rails for technological development that is in its infancy. This is not to say that the philosophy of technology should engage in science fiction, we do not have to elaborate on the ethics of anti-gravity or time machines, as we have no reason to believe that these things are possible. However, if we only waited to analyze, categorize, and evaluate the technology that we can be safe will have a lasting impact on our lives, at least the evaluation if (and if yes, under which conditions) we should welcome these technologies in the first place, will come too late. Thus, an effective philosophy of technology, which is supposed to inform the ethics of technologies in their more specific forms, is necessarily speculative.

If you want to imagine technological progress like a train, philosophers of technology are the land surveyors assessing where the train should go. Whether the train will be able to even get to the places the surveyor assesses may be unclear at the time the surveyor starts their business, as the train has yet to be built and put on its tracks. And maybe the surveyed land, however apt for trains, may turn out to never be laid tracks on, thus rendering the surveyor's work without practical use. However, we would still claim that assessing land when the train is already entering it, probably full steam ahead, is more likely to lead to unwanted results (if not catastrophe) than having sent surveyors out to areas that will never be used.

We can understand the role of a somewhat speculative philosophy of technology in a similar way. In order to even have a chance to have an ethically balanced development and use of technologies, we ought to have thought about these technologies before they become marketable. Arguably, then, with the publications of GPT-3 and BERT and the ensuing consternation of Annette Zimmermann that "if you can do things with words, you can do things with algorithms" (Zimmermann, 2020), and the latest publications of Dall-E (the text-to-art AI, OpenAI, 2022), Gato (the multimodal transformer, DeepMind, 2022), and first sex robots entering the free market, philosophy might not have been early enough in its engagement with trends in computer science.

Lastly, we might even want to be more optimistic about certain analyses: an investigation into the ethics of time machines might help us understand how time structures and influences our planning and decisions. We allow for elaborate thought experiments to function as intuition pumps to inform our ethical decision-making, to motivate some reasons about phenomenology, epistemology, and most other areas of philosophy. Whether an investigation into some technology in its infancy is best thought of as preparation for an inevitable future or a thought experiment must be judged by time—however, both types can provide us with some valuable insight, good arguments, and even design suggestions to face potential challenges and risks and uncover chances for this technology. I believe while our investigation into human–machine friendships is closer to an elaboration of a future in waiting, it would make sense as a thought experiment as well.

2.3   Conclusion

The philosophy of human–machine friendships faces some challenges that we should get out of the way before entering the broader discussion. The burden of proof that such a discussion is not a waste of time and other resources should be taken up with some self-confidence. This self-confidence is justified, as we have seen that the three strong challenges—conceptual, ethical, and empirical-political—can be resolved or ameliorated in such a way that it is up to the features of the theory of human–machine friendships whether it is acceptable. Elaborating on the issue in general, however, did not turn out to be an issue.

Additionally, the often presented argument that the technology itself is neither ready nor likely to be ready to the degree presupposed in such theories could be refuted: if we waited until a technology is ready (rather than guide their readiness ethically), we would face some serious issues if the technology matures to marketability. We may even want to say that usually, technological development is too fast anyway for proper philosophical reflection and thus speculative philosophy of technology is not speculative enough.

REFERENCES

Aristotle. (2000). *Nicomachean ethics* (R. Crisp, Ed. and Trans.). CUP.

Crawford, K. (2021). *Atlas of AI. The real worlds of artificial intelligence*. YUP.

DeepMind. (2022). A generalist agent. https://www.deepmind.com/publicati ons/a-generalist-agent. Last accessed 15 June 2022.

Kempt, H. (2021). Zwischenmenschlichkeit für Maschinen. In A. Strasse, W. Sohst, K. Stepec, & R. Stapelfeldt (Eds.), *KI – Die Große Verheißung*. (pp. 453–470). Xenomoi.

Morozov, E. (2013). *To save everything, click here: The folly of technological solutionism*. PublicAffairs.

OpenAI. (2022). Dall-E 2. https://openai.com/dall-e-2/. Last accessed 15 June 2022.

Robertson, J. (2018). *Robo Sapiens Japanicus. Robots, gender, family, and the Japanese Nation*. UoCP.

Zimmermann, A. (2020). If you can do things with words, you can do things with algorithms. https://dailynous.com/2020/07/30/philosophers-gpt-3/# zimmermann. Last accessed 15 June 2022.

CHAPTER 3

Critical Perspectives on Technology and Friendship

As pointed out in the third challenge of the previous chapter, in order to be able to assess the scope (and potentially the contents) of a theory of human–machine friendships, the conditions under which the actual technology is being developed ought to be reflected, evaluated, and critically incorporated. The concern was that our material world in which these social technologies are developed functions in such a way that it precludes human–machine friendships from reaching any kind of depth. Thus, we might be metaphysically or conceptually able to be friends with machines, but the conditions of production rule out any factual chance.

We answered this challenge by stating that we are capable of imagining a world that is not too far from our own in which these conditions are different, or at least not scope-limiting. However, that argument was merely a proof-of-concept style argument to establish that there is indeed a scope of depth in human–machine friendship that is worth exploring. Knowing that, we now ought to turn to the question of what this scope is, and how it is widened or limited by these conditions.

This requires a critical analysis of the different levels of influence, especially the social, economic, and political considerations that inform certain design choices and design requirements. This is also usually the reason for some AI ethicists to reject investigations into more speculative fields of technology: without acknowledging the power structures and incentives within the AI economy, those speculations are doomed to mislead

© The Author(s), under exclusive license to Springer Nature
Switzerland AG 2022
H. Kempt, *Synthetic Friends*,
https://doi.org/10.1007/978-3-031-13631-3_3

(see for two exemplary analyses of the conditions Crawford, 2021 and Coeckelbergh, 2022). This is because once we consider the context, any philosophical abstraction must turn out to be marketing speech by companies aiming to establish a certain marketable narrative. Such a critical analysis, then, should provide some guiding insights into the degrees of the economic incentives, the political agendas, and social norms of programing and developing social machines a certain way.

3.1   Economic Incentives

The research into social machines, especially social robotics, is funded by a variety of both private and public bodies. Private companies, half-private half-public ventures, military institutions, and private and public universities all invest in the research and development of social robots. The economy behind social robots should be critically analyzed in order to understand the potential hurdles in an ethical analysis of the scope and depth of potential human–machine interactions.

3.1.1   Data-Driven Development

Under the current technological paradigm, the improvements for social robots' artificial intelligence come from accruing, processing, analyzing of data, and training large neural networks with them. Considering the special role of communication in social robots intended to establish friendships with, the linguistic skills of such machines will be the most important aspect. Currently, large language models (LLM) are the base model for creating more specific natural language producers, like chatbots or personal assistants. BERT, GPT-3, and other LLM can be used to program all kinds of "semantic machines", to create machines for almost any kind of linguistic purpose. "What we can do with words, we soon can do with machines" stated Annette Zimmermann as a reaction to the publication of the OpenAI-LLM called GPT-3, the largest language model to date.

The economy of these LLM (Luitse & Denkena, 2021), the environmental concerns notwithstanding (Bender et al., 2021), is based on selling and licensing the model to other companies to develop their specific communication product. Most often, these products remain inside the company (Google's BERT) or are connected to exclusive licensing agreements (OpenAI's GPT-3 and Microsoft), while others are open source

(GPT-2, Facebook's OPT [Heaven, 2022]). For revenue streams to remain constant and develop business models utilizing the prowess of LLMs, most products based on LLMs are gathering user data. Another way to create revenue for LLM-based products is subscription models (Macaulay, 2020) of special ad-free experiences, additional features, or customer support (which does not exclude the possibility that the user data is still not collected and potentially processed in some way).

User data then can be used to create profiles to match targeted ads with, or to sell to other vendors for the same purpose outside that specific machine. Social machines, in this logic, are creating the highest revenue if people keep interacting with them, giving companies an incentive to keep users engaged and attached to those machines rather than solving any kind of short-term problem once. This certainly has consequences for the design choices these companies make, the scientific principles they deploy, and the kinds of users they aim to attract.

Thus, the marketing of social machines using LLMs is not to anyone who is interested but is usually specifically designed to attract users that will be creating higher revenue.

Next to the use-based applications aiming to increase engagement and thus seek users with specific communicative profiles, we can differentiate those applications that are associated with a specific location or device. Personal assistants, for example, are usually associated with specific devices (smartphones, smart TVs, smart speakers), often with an associated specific spot (mostly in homes, potentially at work, in childcare- or eldercare homes). For these machines, their ubiquity, simplicity in use and commands, and growing familiarity is the strategy to generate engagement and thus data. This data is not used within the machine but becomes part of the larger company (Alexa as part of Amazon, Google's personal assistant within Alphabet, etc., with Siri within Apple being an exception). This way, economically, these personal assistants and their device-based existence are merely lured for engaging users in most of their daily lives (Furey & Blue, 2019). Many have experienced the almost impertinent self-inclusion of a personal assistant in a conversation if, by accident, the name of the machine is being uttered (or even something is being said that sounds like the name has been uttered). It seems that many people are getting used to having their conversations analyzed and processed by large companies operating in the background. Partially, one can argue, this is due to the subtle anthropomorphization (see also Chapter 7) of the relevant speech patterns.

3.1.2   Subscription Services, Additional Features

We mentioned, besides the data gathering incentives, other versions of generating revenue with social machines: subscription services and additional features for purchase. In the context of the demanded and required sociality of machines, incentives to create long-lasting social connections between humans and machines may be especially vulnerable to exploitation: if users build relations with a machine but have to "subscribe" to their services, or purchase additional features to unlock some interactive dimensions (in, for example, a gamified setup of relationship-progression with a social machine), the company providing these services could establish a long-lasting customer base.

Subscription services, especially those with an upfront or yearly fee, have the effect of binding customers and incentivizing them to invest emotionally and thus commit to the product, and in extension to the social interactions offered by the social machine. The mechanism is not new, but deserves specific attention in our analysis of the contextual conditions of human–machine relationships, as the manufactured dependency of the user on the subscription to retain access to the product will add an irreducible outside dimension to the potential human–machine relationship. Subscription services preclude the impression that one "owns" the product (or even just access to it).

3.1.3   "Replika" and the Gamification of Social Machines

An example of this kind of gamification is "Replika", a conversational companion AI-bot. It is described in its advertisement as the "World's first AI friend" or as a "virtual friend" that "genuinely cares about you" (Luka, 2021). The business model of the app is then a relevant feature to the possibility of actually becoming anything of a friend: the app generates revenue through the possibility of in-app purchases to customize the looks, living environment, and other features of the "virtual friend". One creates a human avatar and gives the entity a name and certain personality features.

The app-developers at Luka Inc. have an interest in creating intra-app incentives for the user to spend money on the app while at the same time creating an experience that is also merely engaging enough for user data to become worth analyzing. Thus, they offer daily experience-packages, two in-game currencies called "coins" and "gems" ("gems" are purchasable with actual currency, "gems" are purchasable with gems) with which in-game outfits and more interactive features can be unlocked. They also offer a subscription service.

While these mechanisms are not new or really anything specific to the current economic landscape of in-app purchases, the fact that this is supposedly a virtual friend that "genuinely" cares about their human counterpart must strike as morally problematic.

With the rather blatant example of Replika in mind in which microtransactions are cynically aimed at the vulnerability of those people who may aim to build a relationship with their Replika-avatar, we can safely conclude that these conditions most certainly render any kind of meaningful relationship impossible. Even if the suspension of disbelief was to last that the avatar itself has no interest in getting money from us (as their "goal" would merely be to grab and retain our attention or improve our lives), the fact that a subscription with an experience-based progression system and microtransactions are financing any kind of relationship with this entity cannot count as something other than cynical. The fact that people still seek this experience, and might even be happy within the constraints of these conditions, does not mean that they are morally justifiable (we can imagine a lot of people being "happy" in abusive relationships right now as well).

3.1.4 Purchasing Closed Systems

Albeit the market is currently being built on these exploitative strategies, it is worth noting that not all social machine systems are pursuing this strategy. There are companies that sell closed systems, i.e., systems that are not necessarily connected to the internet to constantly improve the user experience by pooled user data. These systems come with an increased upfront cost and usually with both hardware (in form of a physical body) and software (in form of an AI). This business model seems to be more popular with companies in business with other institutions like healthcare insurances or retirement homes, rather than those facing the individual customer. These institutions usually purchase social machines for therapeutic purposes in vulnerable populations like people with dementia, neuroatypical persons, or children. These machines cannot be used for targeted advertising, and gamification to increase engagement will also not cause any benefit to the company directly. A subscription service also seems problematic in this context, as therapeutic uses of technology are usually not conditioned on certain bills being paid constantly besides maintenance. The dedication of these machines to vulnerable groups precludes any kind of problematic economic incentive that would expose their users to nudging efforts to increase and deepen engagement. If these machines are not helping the users they are developed for, there is no point in creating incentives for those machines to be more attention-seeking.

Thus, social machines that operate with closed systems are avoiding the exploitative business models of the data-driven approaches. This is probably due to the purpose of their use, and the legal ramifications that regulate the use of such technology in vulnerable groups. However, this is not to say that closed systems will remain the standard for these uses either as we can anticipate the business model of subscriptions to be offered here instead of those closed systems, potentially creating dependencies of care homes and other services toward the manufacturers. Further, we can also imagine state-driven initiatives that provide closed systems of machine friends of a certain level of complexity for free, leaving out the otherwise problematic context of having to purchase social machines altogether (this is not the case at the moment, however, and thus is not an argument in the debate about the current economic context).

3.1.5   Market Conditions

Lastly, we should acknowledge that these developments have different market conditions which determine the quality, accessibility, customer service, and other dimensions of using social machines. As a fast-growing market in different areas of the world, the pool of capable engineers is comparatively limited, often creating rather homogenous workforces and monopolies (Luitse & Denkena, 2021, 2; Srnicek, 2018). These are most often male-dominated, are situated within a specific technosolutionist educational and corporate culture, and exhibit a disdain for ethical considerations that may prohibit or limit their work, legal and political arguments that would outlaw their goals, and social and psychological input that warn them of their work.

We can see these developments, especially in mature tech companies like Facebook and Google that have begun aggressively lobbying for their own political purposes, including relegating the regulation of artificial intelligence to the companies themselves in form of AI ethics departments instead of substantial AI legislation (see below). These AI ethics departments have thus far been woefully undersupplied, strictly incorporated into the company's narrative, or generally ineffective in making an independent case for or against certain technologies, which as some have assumed very much by design of this setup. The well-reported conflict within Google leading to the firing of Timnit Gebru speaks to that being the case (Hao, 2020). In this context, some also speak of ethics-washing (Wagner, 2018), as the presence of an ethics researcher is suggestive of creating ethically justified machines.

Additionally, we ought not to forget that these LLMs are often only trained in one language, and thus the success of these social machines is limited to the available language models. Smaller languages with limited publicly available data may never, or only in the mediate future, achieve levels of communicative AI necessary for human–machine relationships. This is also fueling certain biased assumptions about the quality and dimension of social machines. Even if machine translation services will be able to translate the in- and output of social machines perfectly, the dominance of the English-speaking conversational norms will be noticeable and relevant to the effectiveness of social machines.

3.2 POLITICAL AGENDAS AND LEGAL REQUIREMENTS

As we have seen with the difference in closed systems to data-driven models of social robots, the political ramifications of how AI can be constructed to become usable in social machines are not free of regulatory influence. We should not pretend that we can interact with any kind of social machine without acknowledging past, current, or future political regulations that condition the technology at hand. Thus, the scope of potential human–machine relationships will in part be determined by the regulations put on digital technologies. To get an idea of what kind of basic regulations are already influencing the construction of social machines, we are taking a look at the general demands of the GDPR (General Data Protection Regulation) as well as its article 17 of the "right to be forgotten".

3.2.1   GDPR and Other Consumer Protection Regulations

One of the examples that may be instructive to assess a rather under-regulated market is the European Union's data protection protocol. The GDPR is supposed to function as a check on big corporations sharing user data without or against the users' will, while also giving users the opportunity to differentiate which data is being shared with whom. Violations of this regulation are being pursued with high fees (CMS.law, 2022). As any website offering services in Europe has to fulfill the regulations of the GDPR, most websites worldwide now ask for the consent of their users to utilize their data through cookies and other user-tracking devices. Despite hindering the ease of using the internet, it exposes the previously mentioned data-driven business models by making the data collection apparent and, to some minor extent, transparent. It is not unlikely that in the near future we may encounter similar proposals that specify or correct the GDPR, by expanding users' ability to retroactively deny consent to data collections and request their deletion. As a matter of fact, Jacob Turner's proposal (Turner, 2019) to create new institutions to oversee the development, implementation, and consequences of autonomously operating technology is a first hint at the potential limits of the current data protection regulations as well as the future debate about those regulations.

A similar example is the "right to be forgotten", which consists of the right to have Google and other search engines delete information associated with you to ensure that your past is not indefinitely traceable. The idea for granting such a right is to break the tech companies' grip on their users' data it is capable of saving. The idea that users should be the ultimate authority of what is being done with their data, even after the data has been collected and stored, is representing the slow but steady development of a "data sovereignty" approach in the European Union. This approach is intended to ensure their citizens' digital traces are within their own authority to control. The term "digital sovereignty" could, however, also be translated into regulations that reserve some autonomy for users. This is currently considered to mark the relocation of power from tech companies to users, but could be used to allow for users, fully informed, to consent to different levels of data sharing. Thus, these new fundamental rights are not necessarily determining the limits of social machines, but provide the tools to clearly and swiftly regulate the context (data gathering and processing) as well as content (how social machines will behave) of this new technology.

Turner's analysis and proposal on how to regulate the development of AI and its implementation put these discussions in a wider legal context and on a stronger footing. He distinguishes between the regulators, the creators, and the creations. From a legal perspective, then, each of these dimensions represents opportunities to limit the scope of human–machine relationships (whether purposefully or as a side-effect of other regulatory decisions).

Turner suggests building institutions before making laws, as the former is required to enforce the latter. At the same time, the rules applicable to the creators of AI must determine certain liability questions that may emerge from using certain products. Depending on how autonomous these technologies become, and how the liability question will be regulatorily answered, the economic viability to build machines that may cause strong liabilities for the creators may put the development of human–machine relationships into question. Especially the more intimate purposes a machine is programmed to be used for, the bigger the current ethical (for an overview, see Tasioulas, 2019) and policy (Turner, 2019) concerns may lead to liability-clauses for emotional damage this machine could cause in their users.

The current forms to circumvent potential liabilities require the creator to inform the user, and the user to consent to, a certain level of uncertainty of what the machine may say or do. Equally, social machines that are advertised as providing mental health benefits still require the user's confirmation that no specific mental crisis is motivating the user to engage with this technology. We can expect increased pressure to specify the regulatory side of social machines in the light of growing applications and, eventually, harms being done by those machines.

3.2.2 Trustworthy AI as an Opening in Regulation

On the other side of the regulatory process stands the AI-ethical debate surrounding the European Union's take on "trustworthy AI". While other countries pursue different AI strategies, the European Union's "High-Level Expert Group" (HLEG) has chosen to emphasize the ability to create AI that is deserving of trusted relationships with human users (HLEG, 2019). The idea here is to emphasize the ethical quality of the data gathered and the AI created on such data, thus not only leaving the content of such AI to be trustworthy but also the context in which it was created. The choice of terminology and its implications to human–machine relationships have been discussed at length (see Ryan, 2020 for an overview of the fundamental problem), and it does not seem necessary at this stage to go into this debate. However, providing a list of requirements of trustworthiness in order to bestow that label on some AI opens the opportunity to clearly define to political conditions for allowing human–machine relationships. Reasonably, if an AI is deserving of the politically determined label "trustworthy", then using such AI in human–machine relationship contexts cannot be rejected on the grounds of the technology being limited through legal regulations elsewhere. It creates ethical precedence.

This means, that whatever the label "trustworthy" AI implies, it defines the currently foreseeable scope of human–machine relationships from a political side. Trustworthiness, however, requires the reduction and prevention of bias, as well as quality control to avoid emergent biases (Friedman & Nissenbaum, 1996). It is somewhat doubtful that the EU conditions for trustworthy AI will lead to an effective institutionalization of those ethical criteria. These ethical principles received plenty of industry-backed input and academic pushback (Veale, 2020). The worry here is that this document leaves too many openings for AI developments that are decidedly not trustworthy in any substantial sense of the term, but rather provides ethical cover for otherwise problematic applications (cf. "ethics-washing" as a growing issue in AI ethics).

Lastly, whether these ethical criteria will be translated into codified law will depend on further developments and successes of EU-based AI. As China and the USA are taking different steps toward a normative approach to AI, the idea of an ethically justified AI "Made in EU" seems, thus far, like a proposal only (Abramo & Campbell-Mohn, 2021). In any case, we currently have no reason to believe that these recommendations by the HLEG have any lasting effects on the development of social machines. And, if at all, then in Europe. We do, however unfortunately, have reason to believe that some AI-regulators in other parts of the world will take a much more unrestricted approach to AI.

3.3   SOCIAL NORMS AND PRODUCTION BIASES

The society we live in is both informing the specific contents of designs of social machines as well as functions as a precondition to the technological method of creating AI. The former inspires the product in the first place, and the latter conditions the technology and the way it can behave. Both will influence how we can interact with machines. To analyze both, we can call the former "social norms", representing the explicit or implicit structural norms of our societies, and the later "production biases", in which the data, workforce, and other elements of the production can be discussed as influencing our ability to relate to machines.

3.3.1   Social Norms

We live, generally speaking, in an unequal and unjust world. Historically grown, structurally perpetuated injustices, highly unequal distributions of resources and power, oppression and wars, and silencing and suppression of human rights are occurring daily around the world. We have reason to assume that to some degree, even comparatively more just places around the world harbor structural injustices and always remain at risk of falling back into states of less tolerance. These cultures of oppression, albeit to a different degree, are also present in tech companies. Even though some of the biggest companies in digital technologies promote egalitarian ideals, there is plenty of evidence that subtle forms of racism, sexism, xenophobia, and homophobia are occurring there as well.

In attempting to create social machines, and in recreating sociality, we have to assume that these issues will re-emerge implicitly. Simply by gendering a social machine, we determine certain associations and expectations with the entity. The same applies to race, sexual orientation, and other structural features. This is not to say that we cannot create machines that do not reflect social stereotypes, but often it is hard to gauge what a social stereotype even is. Take, as an example, the issue of using a female-gendered voice as the default for personal assistants on phones. This is not a technical or production bias (as will be discussed in the next chapter), but a specific design choice partially based on social conventions and norms. The subservient, secretary-style role of the personal assistant is apparently well-suited to default to a female voice, as women are still associated with these traits. UNESCO pointed out the danger of reinforcing these stereotypes which inevitably harm women worldwide (UNESCO,

2019). The authors are especially noting that the attempts at sexually harassing the personal assistant with inappropriate comments or questions are manifold, and the responses of the assistants are often koy or even charmed. This suggests, at best, that tech companies do not want to upset their male customers usually making these comments to feel called out for their normally inappropriate behavior. At worst, they are encouraging this behavior as an acceptable standard.

We have reason to believe that an unreflected recreation of human sociality in social machines will lead to a perpetuation of the inequalities currently found in society. And while some societies are witnessing strides toward gender equality, other nations may cement their gender biases by having their own versions of social machines being even more representative of biases.

Conversely, if we overcompensate to rid ourselves of these biases, we may limit the scope of relatability to a degree at which most people do not want to engage (see for such an example the gender-neutral personal assistant "Q" [Mortada, 2019]). This balance, between relatability and anthropomorphism will be elaborated in Chapter 7, as this pertains to the normativity of human–machine relationships rather than the mere ability to relate under real-life conditions in the first place. Depending on the weighing of these normative concerns, we can determine what a synthetic friend can represent.

However, while the implementation—either explicitly or implicitly—can limit the scope of how humans can relate to machines, this does not seem too big a concern for the relatability of machines in general. The fact that they exhibit some form of social bias becomes an issue in their permissibility for ubiquity but not their relatability. If biases and social hierarchies were a problem for social relatability, none of our current friendships could count as such. Being social equals with others would be impossible if an egalitarian society was a precondition.

3.3.2   Production Bias and Technology

Different issues of social norms can be observed by the implicit inclusion of those over the course of the production of a machine. One of the main concerns here is that a machine will not only reproduce some kind of social bias but rather the specific will of those creating the machine and their ideas of friendship. Considering the relative demographic homogeny of the workforce (as mentioned above), we should worry about the subtle interpretations of what sociality is supposed to be, and what it means to be in a friendship in general. Homogeny may be helpful to form a well-functioning team to create software, but it harbors the risks of simplifying otherwise highly complex social circumstances.

The notion of friendship among homogenous groups is culturally embedded, and even when reflected upon not necessarily caught or implementable. Considering now that startups are often comprised of a rather homogenous group (bigger tech companies are, unfortunately not much different in the relevant departments), leading to racial or sexist bias (see, e.g., the study on "the elephant in the valley" about sexism-experiences of women in tech [Women in Tech, 2015]), we should be concerned about the ability of a company to create models of friendship that are not reflective of these biases. Take, for example, any kind of gendered personal assistant and the associated gender-related issues with it. While they are probably not by design, it stands to assume that we will encounter similar effects when creating friendship machines by such homogenous groups. Thus, we should remain aware of the concern about production biases not only based on the potentially biased data but also based on the potentially biased idea of that a friendship, and more specifically a human–machine friendship can or should look like.

The example of Replika that we discussed above is instructive here, as we see a first approximation of what a group of engineers presupposes to be friendship. While we can grant that some of it are due to the economic circumstances of a startup (also discussed above), it seems unlikely that the design of those machines, their way of relating, and the options of interaction are all just due to these economic circumstances. Those who created Replika had a specific idea of friendship, of human–machine friendship, and worked on implementing that idea. However, this point should be more of an encouraging sign to become more engaged as philosophers rather than pessimistic about human–machine friendships altogether. The fact that there are potentially problematic presuppositions about what a machine friendship should do or how they should behave, is instructive to provide engineers with a better theory. Instead of prohibiting human–machine friendships on the count of "impossibility" to achieve the goal worthy of the name, we should then aim at providing a workable theory of friendship that states some necessities rather than rejecting a development that cannot be stopped.

This means, however, an intensified collaboration between engineers on the one side and philosophers, sociologists, psychologists, and potentially even artists on the other. And while this may sound like a challenge for engineers to overcome, considering the challenges an investigation into human–machine friendship faces in the current climate of the humanities, we should expect that philosophers will have a harder time accepting that some of our social connections could receive an automated alternative.

3.4   Conclusion: What About All Three at Once?

When considering that there are hard conditions in all three areas of consideration, we may be inclined to be pessimistic about a company attempting to fulfill these criteria all at once. If a social machine is supposed to be free of bias, economically viable, and meet strong consent-requirements of the user and the political system it is operating in, we may not see how such a machine could be made in the current context.

The economic incentives to produce a social machine with the capacity to become deeply integrated into a person's life suggest exploitative business models of either giving up one's data (given the intimate context of the uses of these machines this brings issues even with consent), or subscribing to services the machines provides, thus establishing highly dubious dependencies that can harm a person severely if they are ended against a person's will.

The political landscape on the one side is currently under-regulated, allowing for high speed of (potentially reckless) innovation in the market of social robotics and machines. With Turner, we have seen that we have to build up institutions to regulate potential codifications of current AI-ethical principles. The HLEG's suggestions of "trustworthy AI" are a marker here, but without evaluating the contents of this ethical rulebook, we noted that these are neither binding nor are the necessary institutions available to enforce them nor do these rules likely be extendable to AI-powered devices outside of the European Union.

Lastly, the social norms and production challenges limit the way we can relate to AI in a different way. While the recreation of sociality brings established power hierarchies into the realm of social robotics, the rejection of sociality would have a limiting effect on the relatability of such machines. On the other hand, the factual construction contexts can have a limiting effect on the relatability of machines: with a limited pool of engineers, their often well-intended attempts in improving the social biases encoded in data can be harmful or futile. The often toothless ethical assessment of such technologies, coming from within the companies producing these technologies, is usually not amounting to the strong guidelines necessary for those machines, while philosophical assessments from the outside often remain without any influence.

All these reasons suggest that the context for genuinely relatable machines (granted that we can make them work) is factually absent. Does this mean we should not pursue an investigation into the philosophical conditions for human–machine friendships? I do not think so. Besides the pessimism that these conditions inspire, we have seen that some current uses for social machines avoid the issues: closed systems may be less powerful at the moment because the algorithm cannot learn from all users at once, but they provide economic security required to carelessly approach a machine from a purely social stance.

Further, at least within the European Union, we may see technology emerging that fulfills most of the criteria to be called "trustworthy AI", as the strong data protection laws, the EU's ability to establish and strengthen enforcer-institutions, and a relatively well-educated populace may provide adequate grounds for developing a large language model that can largely avoid biased, harmful, or exploitative effects on its users. Considering the strong public technology research sector, it should also not be ruled out that some public–private partnerships could be cooperating in laying the groundwork for data-collection practices that would support the criteria for social machines.

Thus, we may be pessimistic about the current development but not because they are necessary. We simply cannot rule out the chance that in a few years we may encounter technology that by any reasonable standard has been deemed "trustworthy" and, at the same time, is sufficiently relatable for human users to build heartfelt, strong, lasting bonds with them. How these conditions will look like from a positive perspective, i.e., what must be present for social machines to emerge that are socially appropriate and ethically justified, cannot be specified here. We should feel confident, however, in going into the normative analysis of the concept of friendship to assess whether we will be able to be friends with machines.

References

Abramo, T., & Campbell-Mohn, E. (2021). The forgotten third: A comparison of China, US, and the European Union's AI development. *Journal of Student Research, 10*(3). https://doi.org/10.47611/jsrhs.v10i3.1762

Bender, E. M., Gebru, T., McMillan-Major, A., & Shmitchell, S. (2021). On the dangers of stochastic parrots: Can language models be too big? In *Proceedings of the 2021 ACM Conference on Fairness, Accountability, and Transparency (FAccT '21)* (pp. 610–623). Association for Computing Machinery. https://doi.org/10.1145/3442188.3445922

CMS.law. (2022). GDPR Enforcement Tracker. https://www.enforcementtracker.com. Last accessed 15 June 2022.

Coeckelbergh, M. (2022). *The political philosophy of AI*. Polity.

Crawford, K. (2021): *Atlas of AI. The real worlds of artificial intelligence*. YUP.

Friedman, B., & Nissenbaum, H. (1996). Bias in computer systems. *ACM Transactions on Information Systems, 14*(3), 330–347.

Furey, E., & Blue, J. (2019). Can I trust her? Intelligent personal assistants and GDPR. In *International Symposium on Networks, Computers and Communications (ISNCC)* (pp. 1–6). https://doi.org/10.1109/ISNCC.2019.8909098

Hao, K. (2020). "I started crying": inside Timnit Gebru's last days at Google. https://www.technologyreview.com/2020/12/16/1014634/google-ai-ethics-lead-timnit-gebru-tells-story/. Last accessed 15 June 2022.

Heaven, W. D. (2022). Meta has built a massive new language AI—and it's giving it away for free. https://www.technologyreview.com/2022/05/03/1051691/meta-ai-large-language-model-gpt3-ethics-huggingface-transparency/. Last accessed 15 June 2022.

HLEG AI. (2019). Ethics guidelines for trustworthy AI. High-Level Expert Group on Artificial Intelligence.

Luitse, D., & Denkena, W. (2021). The great transformer: Examining the role of large language models in the political economy of AI. *Big Data & Society, 8*(2). https://doi.org/10.1177/20539517211047734

Luka. (2021). Building a compassionate AI friend. https://blog.replika.com/posts/building-a-compassionate-ai-friend. Last accessed 15 June 2022.

Macaulay, T. (2020). OpenAI reveals the pricing plans for its API—and it ain't cheap. *The Next Web*. https://thenextweb.com/news/openai-reveals-the-pricing-plans-for-its-api-and-it-aint-cheap. Last accessed 15 June 2022.

Mortada, D. (2019). Meet Q, The gender-neutral voice assistant. https://www.npr.org/2019/03/21/705395100/meet-q-the-gender-neutral-voice-assistant?t=1655291657420. Last accessed 15 June 2022.

Ryan, M. (2020). In AI we trust: Ethics, artificial intelligence, and reliability. *Science and Engineering Ethics, 26*, 2749–2767. https://doi.org/10.1007/s11948-020-00228-y

Srnicek, N. (2018). Platform monopolies and the political economy of AI. In J. McDonnell (Ed.), *Economics for the many* (pp. 153–163). Verso.

Tasioulas, J. (2019). First steps towards an ethics of robots and artificial intelligence. *Journal of Practical Ethics, 7*(1), 49–83. http://www.jpe.ox.ac.uk/wp-content/uploads/2019/06/Tasioulas.pdf

Turner, J. (2019). *Robot rules. Regulating artificial intelligence.* Springer International.

UNESCO. (2019). I'd blush if I could: Closing gender divides in digital skills through education. https://unesdoc.unesco.org/ark:/48223/pf0000 367416.page=1. Last accessed 15 June 2022.

Veale, M. (2020). A critical take on the policy recommendations of the EU high-level expert group on artificial intelligence. *European Journal of Risk Regulation, 11*(1), e1. https://doi.org/10.1017/err.2019.65

Wagner, B. (2018). Ethics as an escape from regulation. From "ethics-washing" to ethics-shopping? In E. Bayamlioglu, I. Baraliuc, L. A. W. Janssens, & M. Hildebrandt (Eds.), *Being profiled: Cogitas ergo sum: 10 years of profiling the European Citizen* (pp. 84–89). Amsterdam University Press.

Women in Tech. (2015). The elephant in the valley. https://www.elephantinth evalley.com. Last accessed 15 June 2022.

CHAPTER 4

Social Philosophy of Technology

We thus far have seen that an investigation into human–machine friendships is not doomed from the start despite some issues with current tech companies' business models, political regulations, and general social norms. With such a somewhat optimistic outlook we can now turn toward some more general considerations of what it means to relate to someone, and then what it means to relate to something. Thus, we first turn toward the very idea of relational philosophy, to then turn to the relational philosophy of technology.

Relational philosophy is somewhat of an awkward term for different paradigms and basic assumptions, as it might suggest a certain historical development. Gunkel, for example, speaks of the "relational turn" in the philosophy of technology. However, relational philosophy as a contemporary category should not ignore the fact that pragmatic philosophy has been exploring several of the dimensions relevant to relational philosophy before and thus should be included as an early way to recenter human beings within their practical horizons, including social ones.

The previously mentioned term "relational turn" is especially coined for the relational philosophy of technology, which has seen a rise in attention through the publications of David Gunkel and Mark Coeckelbergh (among others). These will also form the main points of discussion on this issue. However, especially in the growing debate surrounding the relatability of artificial intelligence (mostly growing due to the widespread

H. Kempt, *Synthetic Friends*, https://doi.org/10.1007/978-3-031-13631-3_4

misunderstandings in the conceptualizations of what makes AI in the first place) will feature other scholars as well—human–machine communications scholars, sociologist, psychologists, even theologists discover the relevance of autonomous technology for their studies: the relationships humans are forming (or are reasonably expected to form eventually) ought to be understood from these perspectives not merely to enable engineers to go even further but rather to have humanities catch up with these developments, provide guidance, ethical insight, risk assessments, and, if necessary, lobby for specific design choices or legal standards and restrictions.

Only if we understand what technology we are dealing with, how relational philosophy and especially relational philosophy of technology have been involved in struggling to understand the nature of our relationships of all kinds, and how these two come together in topics of the moral standing of machines, human–machine friendships, romantic contacts across technologies, and their overall impact and looming ubiquity—only then we can form educated opinions on how to proceed in our technological progress with social machines. The first step toward this understanding is intended to be provided in this chapter.

4.1   Social Philosophy—Relational Philosophy

It should be fair to say that philosophy has been concerned with social issues from the very inception and throughout its history. Aristotle can count as the main philosopher to structurally explore the conditions of human society in a descriptive sense (rather than Plato's idealized society). Famously, he diagnosed human beings to be zoa politikoa, beings of the polis, and thus laying the groundwork for incorporating sociality into the human condition.

This first characterization of the human condition as one fundamentally social should not be underestimated in its importance for philosophical thinking and has enjoyed over the few decades an ever more growing incorporation to other fields of philosophy: social epistemology points out the inherent social and cooperative nature of knowledge, feminist philosophy is establishing the idea of relational autonomy as basis for agency, relational egalitarianism presents a theory of justice requiring social relations to be largely equal to be acceptable. We may even want to see Wittgenstein's private language argument as pushing philosophy of language toward social considerations.

This goes to show that philosophy is performing a however slow "relational turn" (Gunkel, 2022).

Philosophical inquiries into features of the individual have for a long time been limited to investigating the limits of the individual person. The "relational turn" is one instance, especially in the philosophy of technology, that points toward an individual's connectedness and willingness to connect.

Another branch of philosophy concerning itself to recenter individuals as socially embedded beings is feminist philosophy. Feminist philosophy, as we understand it, premises its thinking about human beings on a simple yet wide-reaching premise: that humans are connected to each other, in their constitution and fundamental properties, in their beliefs and preferences, and in their decisions and actions. Thus, in order to understand the nature of human behavior and mental processes, an analysis must include the social connections that enable a person to not only create but also use the social connections that provide them with morally relevant features. One of the main examples discussed in the literature is relational autonomy and the insight that for individual decisional ability, we must account for the social contexts and connections a person has (see, e.g., Nagel & Reiner, 2013; Walter & Ross, 2014).

4.1.1   Othering

One last important feature of philosophical concern, coming more into the focus of philosophical analysis in the last decades, is the concept of the "other" and the process of "othering". The "other" takes its roots from phenomenological analysis in which the perception of other minds and the shared intersubjectivity of the world. Thus, within phenomenological analysis (coming from Hegel, Husserl, and de Beauvoir), it is a useful tool to capture the principal difference between one's personal, familiar subjectivity and the observation of someone else's subjectivity to which we have no access but our own perception of their behavior.

The more applied and politically relevant context of the concept, however, emerges when we apply this process on a larger scale within a social context. "Othering" describes the process of distinguishing between groups of people based on certain properties.

The othering of a certain group of people (or individuals belonging to that group) as "different" based on certain properties has been used to disenfranchise these people to participate in public and moral discourse or even just counting as part of the moral community (Brons, 2015). Othering, in this reading, is an often problematic occurrence of creating a somewhat privileged or elevated "ingroup" of those who belong (to a space, a culture, a context) and an "outgroup" of those who do not belong (or whose belonging has to be justified or argued for). Othering, thus, occurs in considering certain default assumptions about who is part of a certain group, or in straightforward attempts in downplaying similarities between an ingroup and an outgroup, e.g., through dehumanization, exoticization, and vilifying (Brons, 2015). Unsurprisingly, the decision of who belongs and who does not is usually made by the ingroups, rendering "othering" a process of exerting power of other groups.

4.1.2   Conclusion

These points are all intended to demonstrate that an investigation into the conditions of human activity should focus on the relational dimensions of our existence. We are not monads that choose to extend our hand toward others to make a connection with them. This picture has too long dominated the discourse on certain properties of personhood (agency, autonomy) and suggested that once we figure out individual agency, we would be able to reconstruct sociality according to individual actions and preferences. The opposite is a better picture to understand sociality: social connections are constitutive of these properties, of our beliefs, and our moral sensibilities. Yet, the dark side of sociality is that we tend to think in ingroups and outgroups of those who belong and those who do not. Relational philosophy ought to reckon with the fact that we both are deeply embedded social creatures, and yet seem to privilege our own, and our ingroup's, perspective over others.

4.2   Terminology of Social Technology

The established term for relatable technology on a complex level and thus an immediate candidate for using in this book as well is "social robots", and the discipline developing them (social robotics), with another term, "social technologies", also being used. However, we are going a bit of a different way. Not because there is something wrong with these terms, but they are in itself helpful descriptive markers for certain kinds of technologies. Social robots denote embodied, immediate, and interactive entities and thus constitute a distinct discipline both in research and in development. Psychologically and from an engineering and material sciences perspective, a bipedal robot impresses us differently than one that is quadrupedal or levitating, motivating the concentration on social robotics as research subjects.

However, not all technologies we will consider here have to be embodied to count as worthy of consideration for friendship. As a matter of fact, our primary concern lies with the linguistic capacities, as, as we will in a later chapter, this appears more relevant for what a friendship should provide (though, of course, we can imagine mainly nonverbal, mostly physical human–machine interactions as well).

Thus, the term "robots" is rather limiting or misleading. This issue is also demonstrated in the robot rights debate, where the perspective on embodiment may be misleading the debate to a degree (Kempt, 2020, 161). A different terminological choice could be "social technology", guaranteeing that any kind of interactive technology is included in these considerations. However, this may be over-inclusive. While some approaches to human–machine friendships stress the importance of social media as an additional dimension to friendship in any form (Elder, 2018), thus justifying "social technology" as an umbrella term for some investigations. In this investigation, I invite the reader to use the term "social machine". Three reasons speak for this choice.

First, it is more precise for what we are trying to do here. A social machine is an individuated but not necessarily embodied entity with which we can interact. In opposite to "social robot", it is thus more inclusive of chatbots and personal assistants, while still being specific interactive entities that cannot be considered mere "social technology" that also includes mediums for human–human interactions. Second, it is not uncommon or overly specific. Social machines do not challenge current terminological traditions or come with historical ballast on how parts of the term have been used before (see for example Hendler & Mulhevill, 2016). A social machine, thus, is a clear term to capture that we can engage with technology on a personal, relational basis that works predominantly via verbal or textual communication and optional physical contact (as we will argue for the possibility of friendship with unembodied machines in Chapter 7).

Third, even if this term for this specific set of social technologies is not being taken on, the use of "social machines" is not a challenge for incorporating this investigation into the broader circle of the opus on human–machine relationships. While the term has advantages in use over others, it is not establishing incomprehensible hurdles in utilizing this theory in other terminological contexts, for example, in human–robot interaction studies.

4.2.1   Why Are Social Machines Relevant

While the Chapters 1 and 2 have already provided some implicit comments on the technical issues of social machines, the philosophically interesting and challenging dimension of this technology has yet to be discussed. Most philosophers can agree on the fact that these machines will create ethical conundrums: they may be risky, the mediated influence of those creating these machines on those using them should be made transparent and justified, data protection and transparency should be guaranteed, while the explainability of their inner workings elucidated to a degree in which questions of liability and the responsibility gap (Matthias, 2004) can be discussed productively.

What philosophers cannot agree on is how to deal with and evaluate the phenomenological dimension of these human–machine interactions. Designed as social machines, these devices we interact with exhibit certain social features, incentivize and trigger certain behaviors of humans toward them, and it seems, even if we not incentivized or triggered by those machines to treat them in a certain way, many people still do. From the rather famous examples of bomb-detecting robots being anthropomorphized to become "part of the group" (Carpenter, 2015) to stories of cleaning robots committing "suicide" (Pocklington, 2013), it seems that some people enjoy treating autonomous, interactive machines with a certain drive toward at least social awareness.

Lastly, people will make large claims about the machine they relate to for which we should have an explanation. Not a psychological explanation of why some people make these claims, but a philosophical explanation of what the ontological and social implications of such claims are. We might dismiss some people's claims about their sexrobot being of its own character as someone mistaken about the nature of their device, though such a dismissal requires some more elaborate ethical consideration (see for an overview Nyholm & Frank, 2020). However, the latest (and certainly not last) instance of a Google engineer claiming his sincere belief that the chatbot he has been working with, LaMDA, has developed some sort of sentience (albeit quickly criticized by a large group of engineers,

AI philosophers, and others), should point toward a direction of technology that necessitates a philosophical theory of how to deal with these machines socially and morally (Coeckelbergh, 2022).

Some theories of how we can understand relatable social machines have been put forward that all aim for a similar goal, which is to detach social or moral standing (or "moral patency") from ontological features such as "participating in personhood". While this program seems rather uncontroversial and given the technological developments and some human's willingness to include autonomous machines into (at least their own) social and moral considerations, it still requires an assessment of the consequences of these choices.

In the following, we discuss three of these relational approaches, to then infer some criteria when we want to talk of social machines. In doing so, we avoid having to consider para-social relationships that may not be considered part of such an investigation. However, we do consider and discuss some of the criticisms leveled against these kinds of approaches, that echo the concerns from Chapters 1 and 2.

4.3   David Gunkel and Moral Patency

The current debate surrounding relational technology would not be possible without Gunkel's work (Gunkel, 2012, 2018, 2020, 2022), since he has been an early advocate for expanding the debate concerning the social and moral standing of robots, has suggested paths of rethinking the way we frame the debate, and provides terminological expansions to enable a debate in the first place. For Gunkel (as well as Coeckelbergh), the question of social standing and of moral standing are intimately connected, to a point where analyzing the one without the other is problematic.

The main departure point is the fact that we face some problems with the properties-approach of the moral consideration. The first one being a concern about which properties should count, the second one being how to define these properties to avoid confusion, and the third one being how to detect and determine those properties to be present in a machine in the first place. When someone is operating from, e.g., the condition that consciousness should be a necessary property for a certain moral status (i.e., moral patency which is the consideration of moral worth without being a moral agent), they thus have first to defend why consciousness is the necessary condition, what they mean with consciousness specifically (i.e., they must provide an agreeable theory of consciousness), and they must show that the thing in question either is or is not conscious for the ascription of moral patiency to be convincing. This is not only epistemically difficult but might also lead to ethical issues, as the determination if something is conscious may be too vague to ever be determined for sure.

4.3.1   Moral Standing Leads to Rights?

Moral patiency takes a key role in Gunkel's approach to assessing these relationships between machines and humans, as granting them a social role in our lives is simply not enough. The moment we make any effort into taking a social stance toward a machine, i.e., treat it as (someone) "other" rather than merely as a thing, this social stance is hollow if we do not, at the same time, consider them worthy of moral consideration. In this sense, social recognition and moral recognition are linked.

Gunkel introduces the term "moral patiency" to capture these relations between humans and machines without having to extend anything anthropomorphic onto machines. He states, with Levinas, that the recognition of someone (or something) as "other" is the precondition for agency and patiency, rather than the other way round (Gunkel, 2012, 175). We thus have no analytic access to these terms without already perceiving the objects in question in a certain way. Here, Gunkel and Levinas reintroduce the "other" in the phenomenological tradition: the problem of other minds, i.e., the problem of the unknowability of consciousness in others, is recast not as the hard problem to begin ethical consideration, but as the opportunity to acknowledge that the cognitive capacities are not the decisive element, as the ethical recognition should be foundational (Gunkel, 2012, 176).

Often enough, the debate of whether machines will ever enter the moral circle is limited to the understanding of what machines can be or ought to be. Personhood, or agency, are rather strict and metaphysically murky ways of granting moral consideration and thus expanding the moral circle toward someone else. However, framing the discussion by the question under which conditions we could for certain declare "personhood" achieved, if there are any such conditions to begin with, will keep the debate tethered to a binary affirmative-or-negative stance. However, we potentially will not be able to tell the difference between something resembling a person of some kind, and an actual person. Insisting, then, that there must be a metaphysical difference must lead to disaster.

Thus, Gunkel's issue, which has been picked up by other thinkers as well, consists in the distinction between reification, i.e., the thorough determination of anything technological as a thing, and personification, in which technological artifacts will be treated on the same level as a human person.

Following the establishment of moral patiency through social recognition, the next step in this expansion of the moral circle to include machines lies in discussing the ability of machines to be the bearer of rights. Gunkel opened the field not only through the question of whether they can but also whether they should (cf. his work on "robot rights", Gunkel, 2018). This is because philosophers, legal scholars, and others disagree on both the question of fit and the question of appropriateness. We might have a specific reason to consider robots "technically" capable of being bearers of rights but account for moral reasons to reject such a proposal. Or we could see that it becomes eventually a moral imperative to extend legal protection to machines qua them being machines even though we want to retain a personhood-based concept of rights.

This allowed the debate to flourish, as the distinction between whether something can and something should have rights is now redistributing the burden of proof for those rejecting robot rights: they need to clarify whether they believe if robots cannot be bearers of rights, and why they should not be even if they could. See, for example, Josh Geller's work expanding the concept to not only robots but other entities (Gellers, 2020), and Joshua Smith's approach to the theological implications of this development (Smith, 2021). At the same time, as Gunkel himself points out, this debate merely proves that our understanding of rights as a normative tool to grant and extend protection and claims to each other requires refinement. The question of whether robots should have moral patiency and rights, thus, informs us at least about the concept of rights.

4.4 Coeckelbergh on Growing Moral Relations

Somewhat in parallel, with a similar direction and purpose, Mark Coeckelbergh has developed his approach to the question of moral standing (Coeckelbergh, 2012, 2014, 2018). For Coeckelbergh, however, the single most important requirement for moral consideration of some technology is its relationship to a human (who already is deserving of moral consideration). Thus, patiency is a transitory feature from those who undeniably deserve moral consideration toward those things that are close and dear to humans what are otherwise controversially debated to have or have not moral standing at all. It is thus an ever-growing moral network of consideration and, in this sense, explains some features of moral progress quite well: both the growth of moral connections between moral agents and some entities, and the recognition of those relationships as the relevant feature for "moral standing" have resulted in these entities to be considered moral patients.

This allows for growth of the moral circle over time and creates what Coeckelbergh calls "social ecology", which describes the rather networked, interconnected social reality we operate in. This reverberates with the insights from feminist philosophy as well.

For both Gunkel and Coeckelbergh the criticism of moral relativism has been brought up (Tollon & Naidoo, 2021). This charge is motivated by the seemingly arbitrary reasons for which we can enter human–machine relationships, and thus can extend moral circles according to any conviction's requirements. Further, the evidence of different handlings of what is considered morally relevant has been brought up in Chapter 1 before, when we discussed why technosolutionism should not be a strong argument against human–machine friendships. The fact that some cultures do in fact seem to include machines into their moral calculus much quicker than others, and given that Coeckelbergh and Gunkel both endorse this process, we might agree that this should lead to a relativistic approach to moral circles. This is because the growth of moral circles through growing relations is not a settled direction. We can easily imagine (or rather, find) cultures in which those relationships to machines will be fully rejected, with no philosophical way for including machines into moral circles. If moral standing is motivated by moral relations, we might remain uneasy with this diagnosis, as we can foresee that moral standing is fleeting if people merely decide differently.

At the same time, we should not take any kind of (claimed) human–machine interaction for granted, though we should differentiate between those relationships that Coeckelbergh deems relevant to investigate further and those that are "mistaken". This difference is relevant as it demarcates those relationships that deserver philosophical attention and those that do not. The differentiation between those is, arguably, the same as the determination between those who count as machine-friendly and those who reject machines as any kind of social entity.

4.4.1   Expanding the Premise

Lately, Virginia Dignum (2022) has put forward the call for directing attention toward a relational interpretation of artificial intelligence, in which she echoes Gunkel and Coeckelbergh. However, her focus is informed by the supposedly rational approaches to AI and thus focuses on the difference between rational and relational understandings of AI (Dignum, 2022, 9).

Rational understandings of AI, in her view, are based on the question of the benefit and purpose of AI, even social ones. Viewing AI as a relational technology, in turn, is geared toward the relevance of the technology to our lives. This is also a way for Dignum to introduce non-Western and non-analytic traditions of philosophy to the approaches to assessing and regulating AI. The two philosophical traditions she identifies as promising are the Ubuntu philosophy and feminist approaches to technology.

Both are built on the understanding that relationships are essential for the development of human beings, for their moral character and ability to make decisions. The Ubuntu-approach centers the community and the consequences of moral decisions on that community (without, as Dignum explains, undermining individual rights). Recognizing this centering, Dignum proposes to incorporate these premises into considerations where the relational context of a human being is being affected by AI-decisions, e.g., medical decision-making and human rights (Dignum, 2022, 10–11). She also acknowledges that for a culturally sensitive implementation of responsible AI-devices especially in the context of its origin, Ubuntu may prove successful.

On the other hand, feminist approaches to technology also propose a relational sensitivity for developing, deploying, and assessing AI. Feminist approaches base their premise on the same perspective as Dignum is suggesting here (Dignum, 2022, 12). This premise consists of the assessment that philosophy has given too much attention to the individual capacities without accounting for the conditions and requirements under which these are formed. Even concepts like autonomy, usually conceptualized as an individual feature of persons, ought to be viewed through a relational lens. In this context, Dignum suggests that transferring this feminist framework can expand the work on relational AI. In her account, feminist approaches can contribute to readjusting the work on ethical AI, from merely solving the problems this technology causes to reorganizing how AI ethics are being done.

As she is excavating the necessity for further investigations, she reinforces the judgment that the "relational turn" as diagnosed by Gunkel is well underway and will become more relevant the more philosophical traditions and schools of thought turn their attention to the relational basis of this technology.

4.4.2   Somewhat Similar Approaches

Gunkel and Coeckelbergh (and others, see, e.g., Darling, 2021; Tollon, 2020) should count as proponents of an essentially very similar proposal: that the relationships we build with machines should count as the key to assigning them moral standing, while we should also reject any idea about moral standing being motivated by a metaphysical feature, like participation in humanness, personhood, or being an earthling. Rather, both agree, should we investigate why people build these kinds of relationships.

However, rejecting any kind of "fundamental" or "grounding" element of such a relational feature must push someone to some unwelcome moral relativism which can haunt their proposal: once we give up metaphysical requirements about personhood or moral participation, we might end up with, technically, anyone or anything being allowed in. While for some deliberations it might make sense in attempting to include as many perspectives as possible, we should hope that, for a good number of moral concerns, the main answers of moral philosophy will hold. Thus, we should seek more than a relativistic "anything goes" for moral concerns (if we count Gunkel's approach to be lacking in terms of substantial additions to the metaethics of discourse participation). In the following, we will present and discuss the idea of "pragmacentrism" which is designed to avoid the charge of moral relativism by reintroducing some strong requirements of who should and counts as deserving of moral consideration. This way we expand Gunkel's and Coeckelbergh's approach to moral consideration without incurring the same issues.

4.5   Pragmacentrism

Kempt has introduced another alternative of relating to machines socially without having to revert to ultimately arbitrary (both personal and cultural) emotional responses. With this approach, we would avoid the relativism-charge against the previous concerns. Thus, we should check whether this answer can get us to where we need to be, philosophically, or whether we require some objective feature after all.

The idea of approaching this without presupposing substantial metaphysical features stems from the philosophical school of Erlanger Constructivism. This approach is aimed at pragmatically reconstructing human activities, including ethics and logic, by relating them to their lifeworld counterparts, i.e., their pre-theoretisized activities. In this way, those constructivists reconstructed ethical discourse as an endeavor for the right thing to do by relating them to the way we resolve conflicting action goals through rule-based negotiation. The benefit of reconstructing ethical discourse this way is the irrelevance of the label "person" or even "human" in such discourse, as the ability to participate in such discourse is the decisive measure for full moral consideration. This approach to moral consideration has been named "pragmacentrism" (Gethmann, 2002; Kempt, 2020).

Kempt argued that machines are clearly predisposed to eventually fully participate in such discourses if concepts like "action" or other requirements for discourse participation are not spelled out to be implying "personhood". From a purely pragmatic perspective, if someone or something can successfully participate in ethical discourse, e.g., by advocating for themselves or by taking any specific moral stance, it becomes borderline impossible to reject their request or claim. This is because pragmacentrism does not require anything but "actions" to become part of the moral calculus (compare here also Danaher's ethical behaviorism [Danaher, 2019], which, depending on the concept of action, can come very close to this idea).

This approach also avoids the metaphysical, looming concern about machines eventually forming personalities and thus requiring the full force of moral consideration: There is little to no difference in the moral status of a person and a discourse participant. If a machine is fully conscious or merely able to take part in a moral discourse, then, is irrelevant: what counts is the machine's ability to argue in a moral discourse for itself and others.

Of course, the lower spectrum of technological sophistication is clearly a different, albeit probably more important question: if a machine is not capable of participating in such a discourse, how could it even gain moral consideration? The idea here is that pragmacentrism, while determining moral consideration through discourse participation, does not limit consideration to those participating. Instead, if the moral community decides to include more entities to be worthy of moral consideration (but potentially without bearing the same obligations), it can so choose to. The standard examples for this move are the recognition of small children and of animals by agents (i.e., humans who do participate) through a tutoring/chaperoning relationship. For such consideration, no further metaphysical property is to be had, though those who decide to add something to moral consideration have to be able to demand it.

The willingness to extend what Gunkel calls moral patiency to machines, then, lies in the hands of agents, and thus in the moral communities. However, while this implies a rather liberal approach to moral communities, the charge made against e.g., Coeckelbergh's account by Tollon and Naidoo (2021) suggests that this might still open the door for moral relativism. For example, this account could also allow for the possibility of a rather exclusive moral community that does not recognize any kind of non-human non-agent as in any way relevant for moral consideration. Basing the conditions of pragmacentrism on rational participation in moral discourse, i.e., following the necessary conversational demands to come to any kind of reliable consensus, will require certain concessions to those who insist on taking in entities to which there are strong emotional bonds.

Ultimately, this approach claims that while specific norms about the moral considerations of certain technologies should be up to the moral communities themselves, there are rational ramifications about recognizing each others' emotional bonds, and thus eventually adding technologies to moral considerations until this technology is capable of participating in such discourse on its own.

4.6   Synthesis

4.6.1   Interactivity, Autonomy, and "Relatability"

In this discussion surrounding the necessary yet not metaphysical features of a machine we can mark three items to fit in each of the presented categories of what makes a machine relatable without assuming too much about the person relating: the first being interactivity, the second autonomy, and the third being some kind of "relatability". In this three-feature approach, we can see that there are some necessary conditions for human–machine relationships that undermine Tollon's and Naidoo's accusation of Coeckelbergh's approach being morally relativistic.

Interactivity seems rather straightforward: first, we require a machine that can react to a persons's input while also producing a response that the interacting human person can react to as well. If a person can meaningfully interact with a machine, as in have its own purposes changed due to the back-and-forth between them and a machine, we have achieved some interactivity of philosophical relevance. While there is a psychological notion of interactivity that requires less than this, we should expect any machine of relevance for social consideration to exhibit some dimension of this feature. If we cannot interact with a machine we otherwise would consider consideration-worthy, we have no further reason to advocate for its addition to the moral circle. A back-and-forth between the technology and humans, even if about rather shallow issues, should be considered an element of relationality.

Autonomy, at the same time, is equally relevant. If a machine is interactive in the sense of "reacting to our input while producing output that can manipulate our own purposes", then we might be inclined to call this machine "autonomous". If it can set its own goals and, once achieved, can set new ones within a program of achieving generalized goals anew, most pragmatic conditions of a machine to be autonomous are fulfilled. Of course, we should not attempt to apply more demanding concepts of autonomy to machines. Once we attempt to apply, e.g., a strong concept of autonomy to these entities, we are already leaving the relational approach to technology. The very idea of reconstructing human–machine relationships is to find out something about our projection of morally relevant features onto machines, not about making metaphysical determinations.

Relatability, however, is a much more difficult concept to pin down. In some sense, we should want "relatability" to mean something between Gunkel's face, Coeckelbergh's relationship, and Kempt's ability to participate in discourse. Finding a precise definition for a machine that fulfills these criteria is difficult. However, we might not need a precise definition of what "relatability" is, [or in the German version "Zugewandtheit" ("turned-towardness" or "affection")], but an application of such turned-towardness. While Gunkel's face might exactly imply that (and Coeckelbergh's relational bonds might merely represent a certain ability to read and react to Gunkel's/Levinas's "face"), the choice might work better here: we all know the feeling if something is generally turned toward us, or whether it is not. Be it a human person that is giving off "bad vibes" or a machine that merely does not give us anything, we should not dictate what is a relatable technology for anyone in particular. While some relational technology might better connect to men or women or children, some others might connect better to someone who has an explicit and elaborate interest in some hobby. Relatabillity, thus, is a feature of machines that incentivizes humans to relate to them.

4.6.2   Counters

It is worth noting that not all authors working on relational technology pursue a program that would lead to anything but the partial recognition of these technologies as social entities. In fact, many authors in the debate are rejecting the idea of technological social entities with any kind of social or moral relevance. Notably Joanna Bryson (2010, 2018), Abeba Birhane and Jelle van Dijk (2020), Luciano Floridi and JW Saunders (2004), and others. All voice their concern, but with different orientations and specific assumptions, though the overall narrative is the same: from a philosophical perspective, it is highly problematic to add machines to the group entities that deserve moral consideration.

Some base their concern on the fact that relational approaches mask the very truth of those machines by reinforcing an (otherwise understandable) false sense of comradery. This is done by elevating these connections to anything philosophically interesting other than a general observation about the human ability and willingness to extend relationships beyond their species (e.g., Bryson). Their position, then, is that machines should not be treated as even having the potential to take up meaningful social positions and that philosophy should unmask the mechanics of such relation-building and thus lead people toward the truth. While Bryson has softened her tone toward the proper place of machines toward us (from her infamous "Robots should be slaves" she has changed the rhetoric of that demand), many will agree with her that considering machines as any other than tools must lead to problematic outcomes where machine interests are weighed against human interests.

Others point toward the general absurdity of this masking process in the face of ongoing injustices due to power imbalances. By elaborating on the social relationality of human–machine interactions rather than these contemporary challenges of socially and environmentally unsustainable production, valuable attention is diverted to problems that are not (yet) of any kind of substantial relevance (Birhane & van Dijk, 2020). In this view, the offense lies in seemingly misprioritizing the ethical relevance of power exploitation that affects real people at this point.

These criticisms are valid and we should take on the burden of proof to show that the relational philosophy of technology is not only not foolish or even harmful but that it actually can provide some good in the world by helping us understand otherwise fundamentally human-centric notions of normativity, such as rights or moral patiency.

I believe those who work on the relational philosophy of technology usually provide sufficient justification for their projects. As we have done in Chapters 1 and 2, we pointed toward the fundamental challenges for any investigation into human–machine friendships, and into relational technologies at large. The way these technologies are constructed, whose data is used, by whom and how in which corporate culture, the legal ramifications and requirements for these machines, and the social ontology implicit in the construction of those machines all determine the potentiality and actuality of social machines. Thus, without such awareness relational technology must be an empty enterprise.

4.7 Conclusion

Social machines have made great advancements, and so has the development of a relational philosophy of technology. After the "relational turn", the insight that we cannot proceed with an account of unverifiable property-based access to our moral and social circle. Not only because this has shown in the history to be an exploitable argument to keep certain groups out of moral consideration altogether but also that we will eventually reach problematic ethical notions of determining which entity, indeed, fulfills these criteria.

We thus go by the relationality of those entities and have found different suggestions about what makes someone or something relatable. With a specific view for technology, several authors provide some approaches to achieve that. From Gunkel's "face" of a machine in which we perceive the machine as an independent, individuated character, to Coeckelbergh's strength of relationship and Kempt's use of pragmacentrism for moral standing, approaches have been put forward on how to conceptualize moral machine patiency besides metaphysics. We settled on some minimal requirements to allow for these things to emerge, with autonomy, interactivity, and a certain relatability that is informed by the aforementioned theories. These requirements will help us limit the question of what can be a machine friend in this investigation. Without such requirements in place, we would not be able to reject some of the more obviously absurd proposals about friendship, like claiming to be friends with a rock.

However, the very idea of discussing moral patiency has been rejected by others due to its supposedly unserious treatment of the technology at hand and the ignorance of the bigger picture of technology construction. We should take these issues into consideration when discussing the conditions for human–machine friendships.

The takeaway for this kind of issue, as we will discuss further in Chapter 8 (The problem with robophobes), is a priority of phenomenology and a request for magnanimity. Those who enter and cherish relationships with machines deserve to be taken seriously in their feelings and preferences. They experience intimacy with something else and assign high value to these connections, considering them relevant to their well-being, and overall happiness.

We can remain very aware of the highly problematic contexts in which social relational technology is created, with its economic incentives that nudge toward unhealthy uses of such technology. However, neither should we pathologize those who build relationships nor should we ignore these developments altogether. We are, thus, called for offering an interpretation of what is going on that takes into account the lived realities of people in those relationships.

References

Birhane, A., & van Dijk, J. (2020). Robot rights? Let's talk about human welfare instead. *Proceedings of the 2020 AAAI/ACM Conference on AI, Ethics, and Society*.

Brons, L. (2015). Othering, an analysis. *Transcience, 6*(1), 69–90.

Bryson, J. J. (2010). Robots should be slaves. In Y. Wilks (Ed.), *Close engagements with artificial companions: Key social, psychological, ethical and design issues* (pp. 63–74). OUP.

Bryson, J. J. (2018). Patiency is not a virtue: The design of intelligent systems and systems of ethics. *Ethics and Information Technology, 20*, 15–26. https://doi.org/10.1007/s10676-018-9448-6

Carpenter, J. (2015). *Culture and human–robot interaction in militarized spaces: A war story*. Taylor & Francis.

Coeckelbergh, M. (2012). *Growing moral relations: Critique of moral status ascriptions*. Palgrave Macmillan.

Coeckelbergh, M. (2014). Robotic appearance and forms of life: A phenomenological-hermeneutical approach to the relation between robotics and culture. In M. Funk, & B. Irrgang (Eds.), *Robots in Germany and Japan*. Peter Lang.

Coeckelbergh, M. (2018). Why care about robots? Empathy, moral standing, and the language of suffering Kairos. *Journal of Philosophy & Science, 20*(1), 141–158. https://doi.org/10.2478/kjps-2018-0007

Coeckelbergh, M. (2022). *The heart is not enough*. https://coeckelbergh.med ium.com/the-heart-is-not-enough-how-the-controversy-about-a-chat-bot-rev eals-the-shaky-foundations-of-576bf6c8e155

Danaher, J. (2019). Welcoming robots into the moral circle: A defence of ethical behaviorism. *Science and Engineering Ethics*, 1–27.

Darling, K. (2021). *The new breed. What our history with animals reveals about our future with robots*. Henry Holt.

Dignum, V. (2022). *Relational artificial intelligence*. https://arxiv.org/abs/2202.07446. Accessed June 15, 2022.

Elder, A. (2018). *Friendships, robots and social media*. Routledge.

Floridi, L., & Saunders, J. W. (2004). On the morality of artificial agents. *Minds and Machines, 14*, 349–379.

Gellers, J. (2020). *Rights for robots. Artificial intelligence, animals, and environmental law*. Routledge.

Gethmann, C. F. (2002). Pragmazentrismus. In A. Eusterschulte, & W. Ingensiep (Eds.), *Philosophie der natürlichen Mitwelt. Grundlagen—Probleme—Perspektiven* (pp. 59–66). Königshausen and Neumann.

Gunkel, D. (2012). *The machine question*. MITP.

Gunkel, D. (2018). *Robot rights*. MITP.

Gunkel, D. (2022). The relational turn. Thinking robots otherwise. In J. Loh, & W. Loh (Eds.), *Social robotics and the good life: The normative side of forming emotional bonds with robots*. Transcript Verlag. https://doi.org/10.2139/ssrn.4099209

Hendler, J., & Mulhevill, A. (2016). *Social machines: The coming collision of artificial intelligence, social networking, and humanity*. Apress.

Kempt, H. (2020). *Chatbots and the domestication of AI*. Springer International.

Matthias, A. (2004). The responsibility gap: Ascribing responsibility for the actions of learning automata. *Ethics and Information Technology, 6*, 175–183. https://doi.org/10.1007/s10676-004-3422-1

Nagel, S. K., & Reiner, P. B. (2013). Autonomy support to foster individuals' flourishing. *The American Journal of Bioethics, 13*(6), 36–37. https://doi.org/10.1080/15265161.2013.781708

Nyholm, S., & Frank L. (2020). From sex robots to love robots. In J. Danaher, & N. McArthur (Eds.), *Robot sex: Social implications and ethical.* MITP.

Pocklington, R. (2013). *World's first robot SUICIDE as family return to find cleaning gadget had turned to ash.* https://www.mirror.co.uk/news/weird-news/worlds-first-robot-suicide-family-2786901. Accessed June 15, 2022.

Smith, J. K. (2021). *Robotic persons.* WestBow.

Tollon, F. (2020). The artificial view: Toward a non-anthropocentric account of moral patiency. *Ethics and Information Technology, 23*(2), 147–155.

Tollon, F., & Naidoo, K. (2021). On and beyond artifacts in moral relations: Accounting for power and violence in Coeckelbergh's social relationism. *AI and Society.* https://doi.org/10.1007/s00146-021-01303-z

Walter, J. K., & Ross, L. F. (2014). Relational autonomy: Moving beyond the limits of isolated individualism. *Pediatrics, 133*(Suppl 1), S16–23. https://doi.org/10.1542/peds.2013-3608D

CHAPTER 5

Concepts and Theories of Friendship

After having accumulated the conditions for both a relational philosophy and a relational approach to technology inspired by David Gunkel and Mark Coeckelbergh (including its limits regarding different dimensions of social philosophy), we are finally equipped to say something about friendship. However, the things we have to say about friendship may appear a bit convoluted or overly inclusive for the rather limited point we seem to be making later on. It seems excessive to revisit philosophical approaches to friendship at large to simply make a statement on the limited question of whether this theory also applies to machines.

However, as we have seen in the previous chapter on social relationability, the question of whether friendship can be viewed independent of our presuppositions about the sociality of human beings in general. Similarly, we have seen that the basic convictions about the ontology of social machines, i.e., how we answer the question of whether a machine can ever be a social entity or not, will have to influence any theory we can put forward that treats machines. If we accept that machines can in fact participate in any kind of sociality, and we have seen that we should accept this view, then we should recalibrate our concept of friendship accordingly. If we do not accept this, then we do not have to recalibrate our convictions, though we have to explain a different set of phenomenological occurrences without having a helping theory at hand.

H. Kempt, *Synthetic Friends*, https://doi.org/10.1007/978-3-031-13631-3_5

Thus, from a perspective of argumentative strategy at least, it seems prudent to provide a theory of friendship that could, potentially, incorporate machines. However, for this to work we should make several distinctions that will be reflected in this chapter. First, we will turn to the rather rich phenomenology of friendship and propose psychological, sociological, and ultimately philosophical approaches to characterize friendship. The differences in the definition of the respective disciplines do not suggest that these disciplines cannot have a shared understanding of what a friendship is, though differences persist. Rather, these differences are reflective of the research interest that guides the different disciplines: psychologically speaking, a friendship constitutes something fundamentally different from what sociology would presuppose. Yet, both psychologists and sociologists can be friends with each other, and even can talk about the same things in their (rather varied) research. The differences in kind and in quality are to be explained along the different *research interests* that motivate investigations in the matter.

These research interests are all legitimate and can inform each other's investigations. We will, in fact, refer to psychological studies about the individual perception of machines in social contexts regularly; we also well incorporate initial studies about the social integration of machines based on social acceptance and reception in our later investigations. However, these comments beg the question of what the research question of a philosophical investigation into friendships is. What can philosophical research provide, in opposite to psychological and sociological research, to a debate on friends? It mainly seems like a normative task.

Normative contributions are immensely valuable for our understanding of a certain subject matter, or our responses to such subject matter. Understanding how anger, or hurt, or happiness play both a role in an individual's life as well in collectives. Understanding how we should react to these emotions, what these emotions give us reason to do, and how we should respond to those failing to respond to these reasons is an open, yet fundamentally important question to have an answer to.

Thus far, we might have justified more than necessary. It appears rather straightforward to argue for philosophy's rightful place among the disciplines when discussing which research can be of use for our purposes. Yet, we still only know that philosophy could provide a normative insight into what this research question is or can be. Therefore: what *is* this research question?

Philosophy, frustratingly, has given a number of answers, most often at the same time, thus masking the fact that these are indeed different answers. These answers involve conceptual approaches, ethical approaches, and, to a degree, moral approaches. In order to keep these answers separate (even when given at the same time or even within the same statement), it is useful to think of these as different dimensions of the question "what is a friend philosophically?".

As we will see later on, this question even opens up to aesthetic considerations, potentially. For now, we will go by conceptual, ethical, and moral dimensions of friendship conceptions. Before we should jump into philosophical concepts of friendship, though, we should first account for the descriptive, operationalized readings of the psychological and sociological disciplines, as these might provide a blueprint for some of our normative demands.

5.1   THE SOCIOLOGY AND PSYCHOLOGY OF FRIENDSHIPS

Psychological research into friendships varies greatly in its premising conceptual framework, as the discipline has never settled on one specific, binding concept or phenomenon of investigation. This does not mean that psychological investigations are pointless, but rather that we should be careful in the differences in what psychologists presume to be friendships. Generally, they distinguish friendships according to the emotional depth of the relationship ("acquaintance", "friend", "close friend", and "best friend"), with increased dependencies and depth of engagement (Hall, 2012). Thus, they avoid a normative distinction in which ones are better or worse. Further, psychologists are interested in how individuals make friends, with distinctions such as "compartmentalizers", "tight-knitters", and "samplers" (McCabe, 2016). These describe degrees of closeness and whether we prefer friend groups ("tight-knitters") or bilateral connections ("samplers").

While in the psychological sciences friendship was analyzed from the perspective of individual experiences, sociology turns toward the social effects of friendships as a connecting force between two otherwise unconnected individuals. Thus, the subject of investigation of friendship is explicitly not the simple interactions between two humans, but the conditions under which these connections occur, the regularities in which these connections unfold, and the consequences that friendships have in building, maintaining, and reinforcing social norms (Allan, 1998). It also is an indication of general social cohesion, as the more widespread friendships emerge (e.g., across classes, ethnicities, religions, etc.), the more cohesive and open the society at hand is (Vela-McConnell, 2017). The more stratified a society, the less likely it is that people from different strati become friends.

The fact that humans make different friendships depending on different factors outside their control, i.e., based on where people are born, what social status they have, and how their psychological precondition seeks friendships should be instructive that any kind of normative account of friendship should reflect these kinds of circumstances. As we will see, the praise of the most excellent friendships as a rare occurrence is not necessarily a reflection of our general difficulties to be moral, but maybe an indication that we pursue inadequate normative conditions for the best possible friendship.

5.2   Philosophy of Friendships: What Makes a Philosophically Meaningful Friendship?

As said previously, the philosophy of friendships is less interested in the descriptive truth about how we either use the term or actually live our friendships specifically. Rather, philosophy is interested in the concept of friendship as a fundamental human connection, as well as the normative conditions and consequences of entering (or leaving) friendships with other people. As we have stated above, these distinctions are not always clear (or rather often actually attached to each other), and thus it is important to keep them apart as much as possible. Thus, we first turn toward the very concept of friendship (with reference to the psychological and sociological definitions as well) to make clear how some logical or rational requirements must apply to call someone a friend without using the word misleadingly. In short, we first want to find out what friendships are, and then look at the theories of what a *good* friend makes.

5.2.1   Conceptual Conditions

The psychology and sociology of how we form friendships can inform is partially about the conceptual preconditions of friendship in practice. Clearly, psychology and sociology have a shared understanding of the general phenomenon of friendship, even if they might not investigate these phenomena with the same purpose or even the same end. They might not even investigate the same friendships.

Yet, as we have seen, they seem to presuppose similar conceptual conditions that are worth spelling out. This should not count for defining friendship in a specific sense, as this would necessarily mean to afford the arrogance to call someone mistaken in their friendship with someone else (unless that is part of the friendship). What we should have in mind is the classical way of semantically characterizing a phenomenon along the larger features while accounting for the huge variability in the use of the term (both culturally and individually), in the contents of the term, and in the history and future of the uses of this term. Thus, we should seek family resemblance of friendships. Family resemblance, in the classic Wittgensteinian sense (Wittgenstein, 1953, §67), allows for two from each other virtually unrelated phenomena to be named the same name, as we might see the general reason why both are named the same way without them fitting under one specific definition. Often enough, family resemblances are decidedly not to be defined under one hard rule to distinguish them from others. Wittgenstein himself proposes that like forming a thread, the strings used for it merely overlap in parts while forming the item we are interested in. Trying to include every string in the definition of the thread, then, merely describes the thread.

Friendship, I believe, is an exemplary case of this idea of the impossibility of defining friendship. Take, for example, the three different phenomena of two soccer clubs having traditionally friendly fanbases, the mutually advantageous inter-species hunting partners of a fox and a badger, and the intimate, even-leveled relationship between a mother and her daughter. In a sense, a fan from one of the soccer clubs can claim that his team and the other have a "team friendship". If we see a video of the fox and the badger hunting together, we may claim them to be "buddies" or friends. And we do not wonder about a daughter claiming that her mother is her best friend. Yet, these are all very different phenomena, and all these appear to be in conflict with some of the standard stories about friendship.

The features of this family resemblance approach then are mostly focusing on those features that may or may not be part of friendship, without having to claim that these are necessary conditions for them (as Smith points out, the Bible alone knows several different terms for friendship (Smith, 2021). The guiding thesis for the search for these features is that the philosophical relevance of certain friendships over others comes in the normative conditions of *good* friendships.

5.2.2   Features of Friendship

Some of the following features have been discussed in different philosophical approaches of friendship as essentially necessary conditions. Depending on how the concepts in these different conditions have been understood, some authors may consider some conditions as parts of the normative conditions instead.

We can first note something that is usually self-evident: friendships are mutual, positive connections. That does not mean that both sides are equal in their appreciation of each other or the shared friendship, but that both would generally recognize each other as friends in some form or another. This suggests that a friendship should consist of at least two aware and consenting agents (whatever those agents may be). This "consent" is not to be understood in the strong sense, but can be understood as a behavioral acknowledgment of the situation: we behave according to the judgment that someone is considered our friend (this behavioral condition is to reduce the communicative demand consent may have, as we cannot inquire about the fox and the badger being "friends"). We can, however, observe that there is an unforced mutuality of benefitting each other.

We can note as a second point, as we have done in the previous point on the psychology of friendships, that we differentiate levels of friendship. From the purely phenomenological approach to friendship, this feature is pervasive: most of us in fact are in friendships with different levels of depth and demand, and all of us can see how these friendships can vary over time. We can sensibly speak of "close friends", "best friends", "old friends", and others, usually meaning slightly different backgrounds, qualities, and behaviors of those friends. Yet, we also have friends in a wider or looser sense, people we like to meet or spent time with without having to judge the quality of those friends.

These levels of friendships come in several differentiation criteria. We distinguish between old friends and recently made new friends, between close friends and best vs. loose or superficial ones, between those we do specific activities with and those who are part of our everyday life, etc. We can see that in these levels of differentiation the fact that friendships can be ranked normatively, too.

A third conceptual feature of friendship is their intimacy or "privacy". This point appears to suggest already a certain level of normativity, yet this is merely an extension of the first point. English common language distinguishes a friendly from a friendship connection: If someone is asked "Are you friends with x?" and the asked person answers "X and I are friendly", this is usually to denote a certain lack of intimacy or privacy in their relationship. The general disposition toward each other is positive, but they had not had the chance yet to become friends or might have others reasons to avoid or delay befriending each other. The intimacy or privacy meant here is akin to a certain familiarity on a personal level. They have shared experiences or know some specifics of each other's lives that most others do not. They have had access to each other's privacy, and thus have a however small stake in maintaining the friendship (cf. Thomas' account of "mutual self-disclosure", Thomas, 2013). This does not mean that they know something embarrassing about the other person, but that they have let each other into parts of their lives that most other people do not get access to.

Helm (2022) points out another, seemingly conceptual element of friendship that, while not exclusive to friendship, is understood as a necessary condition: shared activities. The idea here is reflective of the literature about the shared interests or activities that form friendships (Sherman, 1987). We should question, however, whether this is a necessary condition for friendship and not a normative condition for a good friendship. I can easily imagine two people being in a mutually satisfying friendship with little to no wish or urge to change anything about it even though they merely meet for one specific activity. As Helm admits, the literature has thus been rather silent on what is meant by "shared activity", and it stands to assume that there is no such definition that would be a useful limit between when two people are friends and when they are not. To drive this point home, we can imagine two friends mainly just reading books in each other's presence, without ever wishing to expand or change their relationship. These "book-reading friends" are, in Aristotle's understanding clear pleasure friends, yet they do not seem to share an activity besides being within each other's presence. We might have some intuitions that these are no friends at all, but we can also argue that shared activities are more reflective of a *good* friendship (and thus a normative condition). We will revisit the negation problem of normative and conceptual conditions in this chapter that might help understand how positive definitions of friendship can cause some troubles.

5.2.3   *Normative Conditions*

After having established that some requirements apply to calling someone a friend, we should turn toward different approaches to being a *good* friend, as one of our first questions to answer was "why are some friendships better than others?" If we have a concept about what makes a friend firmly established, then it would be adequate to provide a clear definition of what makes a good friend. In giving an account of good friendships, it is not yet determined whether this is a moral account or not. Nehamas is rather explicit in his rejection of morality as a moral undertaking, while Aristotle and his followers do confirm morality in friendships.

The following analysis is clearly not a complete list but merely a characterization and a selection. It is thus a conscious choice to both limit this investigation and to contrast different approaches with each other. As most philosophers have bought into Aristotle's three-kind distinction which has also been applied to human–machine friendships, it seems reasonable to introduce that particular theory as well. In contrast to Aristotle, we mention Alexander Nehamas, as he rejects the idea of friendship being any particularly moral good but draws similarities with art. Thus, being in friendship should be viewed with the aesthetic variety and understanding we appreciate art with, rather than the strict moral requirements of virtue. Lastly, bringing a wholly different idea about friendship into the discussion, we consider "chosen families" as a way to characterize modern friendships. This is reflective of people's lived realities and provides somewhat of a middle ground between the distinctly moral approach from Aristotle and the distinctly non-moral one from Nehamas.

It is undeniable that some other interesting or important ideas about friendship have been left out here; however, in constructing an approach for human–machine friendships, we find these three to give us a helpful setup to delineate some norms that are helpful in describing a phenomenon that has not yet come to reality.

5.3   ARISTOTLE

The first, most famous, and arguably still most relevant thinker of friendship was Aristotle. In his Nicomachean Ethics, the basic work on the idea of a virtue ethics approach to the good life, he distinguishes his three famous forms of friendship. We made reference to this distinction on the previous part, however, we have to acknowledge that Aristotle was keen on distinguishing these along with their normative value. Thus, his concept of friendship is intimately connected to any recommendation of how we should conduct our friendships. As an upshot: we should always work on leading more and better virtue friendships, both by being a better person ourselves as well as seeking others who might be better than us and, at the same time, making others better by being virtuous to and with them. In this sense, friendship is not simply a welcome fact in life, or an evolutionary imperative, or a useful improvement of our defenses against unfriendly others, but rather a normative obligation to the moral character and virtuous person.

Aristotle distinguishes three forms of friendships. These are utility friends, pleasure friends, and virtue friends. Much has been said about these kinds of friendship. Yet, as these distinctions are still being discussed to this day, we should provide a more detailed summary both about Aristotle's overarching theory of moral character and how this theory finally grips into the way we conduct friendships (a similar program has been elaborated upon by Alexis Elder in Elder (2018, Chapter 8, illustrating that this is neither new nor controversial).

We begin with Aristotle's virtue ethics. In his mind, the good life (eudaemonia) is realized by acting virtuously and in accordance with the moral order. The requirement is to act virtuously, even though there are differentiations between acting and behaving virtuously. A person has four options to act, of which only one is truly virtuous.

If an agent knows what to do, but wants and does the wrong thing, such a person is vicious. They disregard the good thing to do for potentially minor reasons (such as egoistic motives or a certain carelessness for others). Current philosophical discourse might call these agents "reason non-responsive" (Coates & Swenson, 2013) as we cannot see how a person that acts in disregard to reasons for the right thing would even react to the reasons they have.

If an agent knows what to do, and wants to do the right thing, but ends up doing the wrong thing, we can call this person weak-willed. Even though their actions lead to the same outcome as the vicious person, we want to note a difference in their moral character, as they acknowledge that they do the wrong thing. They are, in this sense, cognitively reason-responsive, yet fail to translate this responsiveness into the action required to perform according to these reasons or to realize them.

If an agent knows what the right thing to do is in a situation, but does not want to do such a thing, they cannot be reasonably called virtuous.

This is the case even if the action, ultimately, is of great moral worth. If the agent got merely moved by a bad conscience or by supervision of others or due to fear of repercussions, they cannot count as virtuous.

If an agent, however, knows what the right thing to do is and wants to do this right thing, we can call them virtuous. In Aristotle's view, virtue (or rather, being virtuous) is the sign of a moral character, and thus the alignment of a) the ability to judge what the right thing to do is, b) the actual ability to perform the right thing to do, and c) the willingness to perform the right thing against potentially competing interests. As a "hexis", i.e., a disposition or condition, it is the learned feature of a moral character.

The concept of a moral character is thus central to virtue. Without having formed a moral character, a person will struggle with every difficult action to find the virtuous path and this never be better than the third type of agent we discussed before. Aristotle's answer in building a moral character lies in developing the faculty of phronesis (prudence) and experience. The more prudent a person, the better will they be able to judge the virtuous action, and the more experienced in acting virtuously, the more likely will someone be able to act according to virtue as well. Virtue, in this view, compounds by experience. It is controversial whether the experience is indeed a necessary condition for virtue, as Aristotle's approach clearly has a certain kind of person in mind that fulfills these conditions first and foremost—older, upper-class men. Children, so Aristotle, do not possess the necessary experience to be virtuous, even if they should have the necessary prudence.

Prudence and experience, then, should lead us in becoming virtuous agents. The question still remains how. Prudence should tell us the mean or middle (1106a26–b28) between two extreme options of action. A prudent person who knows how to find the mean between two extreme options of action, thus will be enabled to follow their judgment of the right action. Eventually, experience will enable us to essentially have no decisional gap between judging the right thing to do and simply doing it. A virtuous person, then, is one that can judge a situation and in an instant act in the virtuous way. The way to teach a person to become more virtuous is thus by teaching them how to improve their judgment (through phronesis) and by exposing them to virtuous people (being taught by experienced virtuous people).

Why do we introduce this quick sketch of Aristotle's concept of virtue and virtuous characters? Because it is fundamental to his famous distinction separating three different forms of friendship, which set up the philosophical discourse about friendship for the following 2500 years. As we will see, in his conception of the highest form of friendship, we are given the advice to become virtuous ourselves. Thus, an understanding of virtue is the precondition for such friendship.

5.3.1   Aristotle and the Three Forms of Friendship (Nicomachean Ethics, Book VIII and IX)

Much has been said about Aristotle's three forms of friendship, yet it is worth giving a short overview of those forms and how they are considered to differ. The reason for his three-type distinction lies in his belief that we have three reasons for loving something: either it is good, or it is pleasurable, or it is useful. Since Aristotle believes that friendships are an indispensable part of life and even sought after by the most mighty and most rich, we should assume that they have something to do with love. Thus, we can love friendships for the reason that they are good, that they are pleasurable, or that they are useful.

Aristotle begins his distinction with the characterization of friends that are of use. Utility friendships, as they have since then be called, are constituted in the mutual benefit of being friends with each other. However, the value of this friendship is primarily derived from the utility itself, not from appreciating the friend for being useful. We love someone as a friend because they (or our connections to them) are useful to us, rather than the person itself. This is an unstable connection based on mere mutual benefit and thus at the constant danger of falling apart once the utility decreases. Additionally, a friendship purely based on utility would mean that someone is merely in it for personal benefit, as no other.

The second emerges from the love of the pleasurable. Something is pleasurable to use if it stimulates our faculties. This has been taken to include mental pleasures, like a shared activity that is enjoyable for both but less so alone. This pleasure friendship can and should include dimensions of most people's friendships: we can enjoy and seek the good times we have with our friends. Enjoying each other's presence and seeking this enjoyment (and thus each other's presence) is sometimes good enough a reason to meet in the first place. No need for more elaborate activities or bonding moments or soul-searchings; going to the movies because going alone is less fun is for many a perfectly valid reason to meet.

However, Aristotle warns, same with utility friendships, that seeking merely the pleasure of a friendship is to limit oneself to something that might not last. We should seek more in friendships than merely gratification of any kind, and thus should seek to appreciate each other as friends rather than as a source of pleasure.

Lastly, following the characterization of those two rather unreliable and superficial ways of seeking friendship Aristotle presents the third form, the one corresponding to the good. This type of friendship, in contrast, seeks the good in a friendship which, in reference to the previously introduced virtue theory, combines to a friendship of virtue. We seek not the pleasure or the utility the friendship with a person provides, but the virtue itself. In seeking virtue of a social relationship, one must be virtuous oneself—it is a normative and conceptual condition to enter virtue friendships to be virtuous yourself. If you are not virtuous but interested in the other person for the virtue, then you are merely receiving pleasure or utility from it, thus not living up to the mutual conditionality of friendship. Yet, seeking the good (rather than the utility or pleasure) in friendship enables us to become more virtuous in the first place: we learn to care for someone genuinely without hoping to eventually benefit from their success, we learn to be happy for others, to support others, to grow as moral characters, and, thus, become ourselves more virtuous. This form of friendship is supposed to last much longer than the other two, as this one is predicated on the genuine interest in the other person and their success rather than in the service this person provides.

This can be tricky to keep apart, but consider this example: Amelie has three friends, Uter, Pamela, and Veronica. Uter is her go-to friend for tips on how to brew beer, as Uter is a master brewer and enjoys giving out advice. They get along very well but rarely talk about anything but beer brewing. Pamela is an old friend of Amelie's, as they grew up in the same town. They both developed a flavor for beer, and have begun taste-testing different kinds together. Their pallets differ greatly as Amelie prefers IPA and Pamela loves a stout, yet their disagreements in taste have led to fun conversations and generally entertaining nights. Lastly, Veronica does not drink at all but also does not judge Amelie for liking beer. They talk about Amelie's passion for the drink, its culture, and process, and even though Veronica has no connection to any of that, she appreciates Amelie's passion and interest. Inversely, Amelie is expanding her horizon on other things than beer whenever she talks with Veronica, and learns how to care for Veronica's interests that are not her own. One day, she has a sudden change of heart. Out of nowhere, her pallet completely changes and Amelie has become a wine drinker. She never sees Uter again, and her evenings with Pamela are much less entertaining than they used to be when both were tasting each other's drinks. Only her friendship with Veronica has remained the same, as Veronica is still interested in Amelie's passions, now for wine, and Amelie keeps learning to expand her horizon.

This example illustrates that the connections between Amelie and her friends are conditional on the dimensions of utility, pleasure, and virtue. Virtuous people, thus, have an interest in the other person, their success, and joy in life without profiting from these things themselves. Virtue friends are virtuous with each other and thus improve their own virtuosity. Veronica's interest, and in reverse Amelie's, for each other are predicated on each other by virtue of virtue. Thus, Aristotle recommends to seek people who make you a better person not by the goods they provide to your life but by their character and perspective.

It is important to point out that these are idealized forms of friendship. As Danaher points out later, most friendships have certain elements of the three. Even the highest forms of virtue friendship can cause utility and pleasure for us—and we can very much appreciate these effects next to the virtuous influence of said friendship. It is not wrong to enjoy the pleasure one receives from spending time with a virtuous person and, have the motivation to seek this kind of pleasure, next to becoming a better person. To stay in our example, Veronica might also be a hilarious storyteller, and sometimes Amelie just needs a good laugh and seeks the joy of Veronica's storytelling. This is invalidating the virtue of friendship between them, but illustrates that even those friendships can be of utility or pleasure. Same applies to Pamela, who might also be very truthful in her way of approaching things and thus Amelia and her may find a level of relationship, in which veracity is an element of a virtue friendship.

To recap at this stage, Aristotle's story about friendship is a compelling and succinct approach to categorize friendship connections. Often, friendships are instantiations of a certain kind of love: they are useful, or pleasurable. These can be very dear to our hearts. Yet, the friendship that is not only important to us, but those that last are the ones in which our personality is being taken seriously, and in which we take the other person seriously as well. His advice, then, is not only to be a better person and thus be a better friend by valuing someone as a person but also to actively seek these connections.

5.4   Nehamas: "Friendship as an Aesthetic"

In his work, Alexander Nehamas breaks with this interpretation of love in the Aristotelian conception in several ways. As he diagnoses one of the core mistakes of the Aristotelian conception these days, the underlying concept of love, philia, was not intended to be universal but was limited to the polis.

In this sense, friendship based on philia cannot be understood as a general moral good, as it is inherently particularistic. Nehamas points out that we are treating friends specifically differently, as we would do things for them that we would not do for others. This is illustrated by the rather clear judgment "If I say, These are my friends, I imply, These are not" (Nehamas, 2010, 286), which we can also find in C.S. Lewis's seminal work where he states "Friendships must exclude" (Lewis, 1988, 86). He claims that this is based on the Christian distrust of friendship as a competition to agape, the idea of love for humanity and love for god. Thus, trying to claim friendship to be a predominantly (or exclusively) moral good is to misinterpret what it means to be a friend in the name of an unintended generalization of the goods friendships provide. What virtue friendships represent these days, in this reading, is the love of agape.

As a counter-proposal, Nehamas describes the wide varieties in which friendships unfold, including situations in which we do bad things. Friendships, in this sense, are not just the appreciation of the other person as a virtuous person or as the subject of our moral concern, but as the subject of our admiration and our intention to do good because of our particularist love for them. Admiration can come in a variety of forms, as all friendships are determined by the individual friends' characters and their inner-relational dynamics, and as such are related to art rather than morality (Nehamas, 2016).

What makes Nehamas' account of friendship useful for an investigation into human–machine friendships is on the one hand the lack of moralizing in friendships. Nehamas acknowledges that the variety of how friendships are being conducted, while not only not lending itself to moral analysis, is also rather messy to grasp in the first place. The lack of guidance and playbook for how friendships are conducted is the opportunity for human–machine friendships to grow. In fact, friendship is such a complicated thing to analyze that Nehamas does it in form of art-analyzes.

On the other hand, Nehamas still seems to imply or presuppose that the admiration and love for each other are constitutive and potentially outside of a machine's ability. It is an open question whether the highly interwoven and primarily aesthetic value of friendship can be replicated with a machine in this way. We could see that friendship, the "unalloyed good" (Nehamas, 2016, 187) be reproduced in human–machine relationships as a new form of relation, which previously unknown, enriching, and exciting dimensions, if we get it right.

5.5   "Chosen Family"—Most Modern Friendships

In the last decades, another term has come into fashion to characterize the strengths and normative commitments many people have with their friends. The term "chosen family" means the assignment of quasi-family norms to non-related people in one's life. Often used by queer folks as a consequence of often being disowned by one's own actual family, the term chosen family has gained recognition as a term suitable for many lifestyles, as people move away from their families to work and live in different parts of a country or the world (Levin et al., 2020; Lewin, 1993).

Friends often functionally become the center of one's personal everyday life, just as family members would if one stayed at home. For many people on the margins of society, banding together as a chosen family has brought them safe spaces to flourish while acknowledging that interpersonal conflict can and will occur even among intimate friends, that age and experience matter in relationships and that family-like bonds might be necessary to withstand the social issues the group may face. The term "chosen family", thus, is a term denoting the family-like structure of a group of friends rather than a summary of one person's connections and thus suggests a socially embedded approach to close friendships.

This is in contrast to the typically bilateral approach philosophy has taken so far, as it is predicated on the social integration of several friends into one group. And this is a way for many people to find recognition of their existence with others, and build a home. However, many people's reality does not require a chosen family as they embrace their born family and eventually start their own.

Why, then, bring up chosen families as a theory when it is neither a theory (but rather a lived reality) nor universalizable? It shows that there is a substantial alternative for theories of friendship that do not require (or can even conceptualize) virtue friendships. Arguably, friends within chosen families are helping each other grow as much as Aristotle's virtue friend does, but within such a complex embedded social circumstance that an analysis according to someone's character is a misplaced exercise. People in chosen families are friends and become better people by the virtue of the quality of their loving, intimate connection, not by the purity of their moral character (as a matter of fact, the mistakes one makes are very much part of understanding chosen families and their social structure).

The idea of chosen family, in its particularism, is thus closer to Nehamas' idea of the aesthetic value of friendship, while it is still a strong moral approach. The distribution of duties to care for each other, to treat each other "like family" in the sense of reconstructing a strong moral code suggests that morality is an element of the good in these friendships. The form of love that has been considered typical for friendship, philia, can under these conditions be complemented by other forms of love. While there has yet to be a comprehensive theory of chosen families as a philosophy of friendships, we can see the fundamental difference in its very idea of friendships functioning as quasi-familial bonds.

5.5.1   Interim Conclusion: Three Offers

At this stage of the investigation, we are offered three approaches that demonstrate the differences in understanding friendships as social connections. While we do not deny that there are more approaches, we find these three to cover a wide spectrum of assumptions on what makes a good friend. Thus, these theories have a profoundly normative character, as they prescribe certain ways of conducting relationships of a certain kind.

The first one, Aristotle's standard approach, considers human connections more valuable and lasting if they are based on virtue. Virtue shows different, even in friendship, but has the common goal of valuing the other person by virtue of them being a person. A virtue friendship builds moral character because it is a constant opportunity to become a better person while watching someone else be on the same path. These relationships gain their mutual benefit through each other's moral growth, in opposition to the other friendship types that suggest a direct benefit of utility or pleasure. And while these utility- and pleasure-type friendships are often mixed with virtue, the form to strive for, so Aristotle, is the one in which virtue is the strongest component.

The second one acknowledges that for a personal narrative, for our creation of meaning in life, and our sense of a good life, the way we conduct friendships is a key element. Good friendships, in this view, contribute to these things by reinforcing what we like about ourselves, by building virtues but also opening wholly new levels of experiences, creating the most memorable moments in our lives without us having to justify the moral virtue of these memories. The best friendships, thus, are those that represent a certain life aesthetic, in which moral, aesthetic, and other values harmonize. This approach differs from the first by acknowledging that virtue is not the only thing to strive for in good friendships, as they cannot be planned, and should not be a pursuit with an agenda. Instead, Nehamas suggests seeing friendships as something that is based on the good intentions between two people who are concerned with being friends with each other. He explicitly rejects the fact that some behavior may be characteristic of them, and that friendships are as individualized as the people in them. As elements in the construction of our life narrative, they play an essential role in influencing our self-conceptualization. They thus share more properties with art than with morality (Nehamas, 2010, 288).

It is thus more comprehensive in its appreciation of how and why friendships are conducted while rejecting the idea that these relationships must be considered a moral good.

The third one accounts for the fact that sometimes, friendships are the most important social consequences someone can have and conduct. Based on the idea that relationships are constitutional for personhood, and enabling of character and potential, this approach to friendship assigns them the highest value in a person's life. An analysis of these relationships according to certain character qualities thus appears pointless or mistaken, as ideally the whole of a person's character is being taken into this quasi-family structure. In appreciating that most intimate relationships need not be the ones from childhood, we observe that many people form "responsibility communities" in which they care for each other to the degree only known from families otherwise. Such an approach, thus, is more interested in the embeddedness of such friendships rather than their analyzability.

These triple theories of friendship demonstrate that a simple answer in the tradition of one of the three can ignore how others conduct their friendships. This is because especially for Aristotle, those normative conditions are moral ones, too. He presumes that everyone is striving for eudaemonia, and becoming more virtuous will lead to a eudaemonistic life. Thus, anyone who is not striving for a more virtuous life, to which virtue friends count, is failing in the pursuit of the good life. However, we might also inquire whether Nehamas would recommend living an unaesthetic life on purpose, or whether we should choose a family that is indeed abusive and horrible to us.

All these different accounts of friendships presuppose a certain path to good friendship based on assumptions about life. And it seems that their perspective on friendship (and on life) are partially or completely incommensurable—Aristotle sees friendships as something fundamentally different for the advancement of human flourishing than those do who seek a chosen family to grow in. Feminist approaches might also reject the idea of viewing social connections as dispensable, or as mere elements in building an aesthetic or virtuous life; instead, they might view social connections as a necessary condition for any human being's success.

While Aristotle presumes that friendships can be individuated and discreetly analyzed, feminist approaches reject the very idea. They take friendship to be an unavoidable social occurrence and thus best analyzed in the conditions under which they form, who forms them, and what social roles and assumptions play into these connections. All that while the approach "friendships as contributors to a life narrative" presumes a holistic approach to life.

By now, it has become sufficiently clear that these views differ. Yet, we should reject the challenge to choose one of them to move forward and test whether we can create human–machine friendships with them that will be satisfactory to their requirements. Instead, we might want first to inquire if these rather different approaches still have a single combining assumption or issue that we might confront them with. For this, we only have to ask a rather simple question to see how far we come: what is the difference between "being a bad friend" and "being no friend at all"?

5.6   What Makes a Friend a Bad Friend

Different theories of friendship and especially *good* friendships have been put forward. We discussed the normative conditions under which these theories claim that mere friendships become good friendships, with only one, the traditional Aristotelian picture, offering a thoroughly normative one. While it is undeniable that there are plenty more, Aristotle's approach remains the one most philosophers consider especially relevant and which is used to apply to questions of human–machine relationships.

This might be because it adequately captures the normative dimension of friendship that others do not incorporate.

Yet, having seen that normative constraints, if not even a normative order determine which friendships are the more valuable, have been put forward raises a challenging question: how do these normative constraints fare if they are undercut?

Presumably, friendship is something more static (in opposition to merely a fluid state of connection): we *have* friends and are in friendships not only when we are interacting with the friend. It is a lasting connection, even if merely a pleasure friendship or a utility friendship: as long as both people are interested, benefitting, enjoying, or seeking the friendship to stay alive, or rather as long as they are not explicitly uninterested or rejecting the friendship, it usually is assumed to be an "alive" friendship. Further, even when we disagree with a friend about the right thing to do or are disappointed in them, the friendship is not always automatically over. We often have difficulties distinguishing "being a bad friend" and "being no friend at all". This is because the negation of normative conditions can be interpreted twofold: either the negation of the normative condition amounts to someone being no friend at all nor it amounts to someone being a bad friend. Finding the difference between these two requires additional conditions that are often not spelled out in this normatively demanding sense.

We can call this the "negation problem", as they presuppose that the challenge is to positively define friendship and thus increase the difficulties in characterizing bad friends. Take, for example, Aristotle's view from above. He claims that virtue friendships, i.e., those based on each other's recognition and appreciation for the sake of each other, are the normatively best friendships to lead. What, in turn, then makes Aristotle a bad virtue friend? If I fail to live up to the virtue conditions of such friendship, it seems unclear when I should count as a virtue friend, and when I should not count as a friend at all.

Aristotle has an answer to this, and Alexis Elder takes up his suggestion from a value-theory perspective: a bad friend is one that is not helping the friendship flourish. Aristotle's advice, then, to ending a friendship like that, is the following: If bad can be set straight to become a good friend again, one should not end the friendship but rather attempt to help the friend to become a better person. On the other hand, if such a person cannot be helped, it is "not absurd", so Aristotle, to end the friendship (Aristotle, 2000, 1165b). Thus, even virtue friendship can have an ending. Elder adds that we still might lose something valuable by ending a bad friendship, but should do it anyway (Elder, 2018, 34). The idea here is that before someone becomes "no friend at all", to whom no friendship norms apply, they must have been deemed to be unchangeable and uninterested in the flourishing of the friendship by violating such friendship norms.

This negative condition catches the issue we might have with a friend's actions without ending the friendship: we can ask and inquire and struggle to get them to change; once we have to admit that the other person is not interested in changing and will further perform actions that go against the flourishing of the friendship, the friendship has effectively ended. While we still have to determine whether a person can be changed, it is our obligation as a friend to attempt to salvage a friendship before "releasing" the friendship.

Within the virtue theoretical setup, this works well. Two friends are virtuous, and as long as both remain committed to each other's well-being qua their virtue, these friendships are stable. However, if we try to measure this setup against some real-life challenges and our everyday friendship phenomenology, we might find this answer to be less helpful to determine when a friend is either bad or none at all.

First, suppose two good friends have been in a friendship for a while. While they have come from different cultures, they always found a way of relating to each other in a virtuous way. Lately, to circumstances beyond their immediate control, they have been growing apart. They built a friendship on the basis of deeply caring for each other's well-being, but have come to believe in different ways of achieving these goals. One has become more religious and concerned with certain moral behaviors proposed by a religious authority, while the other has become more hedonistic and explicitly rejects these religious ideas. Both still recognize each other as good persons, but their caring for each other has become mildly offensive. Eventually, the one determines that their best course of action is to not be friends anymore, even though they still care for each other.

Second, suppose a certain "down leveling"-effect similar to the levels of friendship mentioned in the conceptual conditions caused by a decrease in communication. A close, good friend, after not being involved in someone's life, becomes a "mere friend", and eventually the friendship just dissipates or dissolves. The friendship may, thus, end on its own if both people just lose interest in each other's life without any of them ever being "bad". The norm to remain fully committed to the other person's well-being seems to rarely apply in real-life scenarios where a variety of reasons can limit and ultimately end friendships without them really being "ended". However, as one may interject from experience, some of our oldest friends are usually those that have been absent for months or years, and one reconnects on a level as if the communication had never been interrupted. Certain friendships can survive or hibernate through the respective friends' life developments. We would call them friends throughout these times, even if their friendship is inactive. This observation, however, makes things just much more complicated and the negation more applicable: when do we declare a friendship ended even though nothing specifically end-worthy has occurred?

Third, some friendships are highly context-dependent. Take, for example, a friendship built through technological mediation, between someone from Iraq and someone from Germany. While the ability to root for and contribute to each other's well-being is built on access to certain technology, this can be undercut through some faultless issues. Take, for example, a friendship between two people from countries with unstable internet connections. After years of daily chatting and phone calls, one person's internet is cut by the government, essentially upending their conversation and influence on each other's lives. We would not say that their friendship has ended; however, we also have to admit that the friendship has functionally ended, especially if we can add that it is unlikely that they ever speak again. Their relationship, bereaved of its necessary medium to function, has seized to exist in a more pragmatic sense.

The first point touches upon the possibility of faultless ends of friendships, in which both friends are virtuous but have to end their friendship for that very fact. The second point contends that sometimes it is difficult to assess whether friendships are still "alive" in the first place because inactivity is an ambiguous measure for friendships. The third point introduces the challenges of technologically mediated friendships (which we will discuss in the next chapter) and the chance of suddenly abrupt ends to these due to the reliance on those machines.

All three scenarios are illustrative of the negation problem that still affects virtue theories of friendship. How can we explain the end of those friendships from the perspective of virtue friends? It seems that the requirement to end a friendship, as one otherwise must be failing as a virtuous person, limits the ability to interpret what is happening in less-than-perfect friendships. For a friendship to end, thus, one of the friends must be at fault. Our examples suggest otherwise. That friendships end because some friends do bad things is uncontroversial. But that even the best friendships can end due to the course personal development makes it rather difficult to find a convincing normative claim to distinguish "no friends" and "bad friends". Especially in cases in which the "growing apart" or "hibernation phase" of a friendship results in different perspectives on that friendship, these insights can be rather painful or problematic. If one person does not consider the other a friend, while they still do, this must lead to hurt and a difference in judgment and opens up the possibility that something is done in complete ignorance of the other person's well-being without violating any specific norm.

The main objection to this negation problem might be that the negation problem is over-theorizing what it means to be a bad friend or no friend at all. There are many reasons for which friendships end, from very intentional, communicated breakups to unintentional growing apart and unfortunate life events. To expect that there is a single answer is a misconception about the nature of these relationships.

It is certainly true that friendships end for many reasons and in different ways, and thus the negation problem does not apply to every instance: sometimes, the difference between being no friend at all and being a bad friend is simply a lack of opportunity for the former to prove the latter: if I have been considered a friend by someone, but have personally moved on since then, they might request a favor from me that I reject based on the conviction that we are no longer friends. The person might consider me a bad friend instead because a proper friend would do them the requested favor. Arguably, without assuming some severe misunderstandings, no other social relationship has such an undefined ending.

The negation problem is also not arguing that the undefined endings of friendships are reason to believe that we have no way of defining the beginnings of friendships. Rather, we argue that it is odd that we have a hard time distinguishing bad friends from not being friends at all, even though there seems to be a consensus that we can define good friends according to certain features—the virtuous traits, their role in our lives, their membership of our innermost group of social connections.

What the negation problem shows, however, is that presupposing a normative order of friendships creates a problem when these norms apply outside the perfect circumstances of friendship. This question, however, becomes ever more relevant in the attempt to transfer those premises toward human–machine friendships. Often enough, the concern about whether we can be friends with a machine is answered negatively from the perspective of virtue theory (de Graaf, 2016; Elder, 2018) precisely because of the presumption that the positive definition will suffice and that machines fail at fulfilling it. Now that we see that there are ways of being no friend while being a good person, we should wonder whether the definition fulfills what it sets out to do.

5.7   FRIENDSHIP WITH TECHNOLOGY

There has been some debate on human–machine friendships that should be considered at this stage, as they largely connect or contribute to the Aristotelian picture. In the following, we will discuss some of those approaches that aim at building a bridge between this Aristotelian picture and the topic of human–machine friendships. The two main points of this debate are the works by Alexis Elder, John Danaher, and Helen Ryland. As we are going to analyze Danaher's later one, we will concentrate on Elder's and Ryland's approaches. For both, we conclude that something else is needed for our approach to grip the notion of friendship in a way that leads to transferable conditions for human–machine friendship. Either because they actually deny that such a connection is possible (Elder) or because their approach is problematic in some other way (Ryland).

5.7.1   Danaher

John Danaher's much-discussed contribution (2019a) to the debate surrounding the possibility of human–machine virtue friendships has been discussed at length else (Nyholm, 2020; Ryland, 2022). Thus, we can keep the presentation of his approach short.

For a plausible approach to friendships between machines and humans, Danaher presents four pragmatic criteria that must be present in a friendship for it to be considered a virtue friendship. In contrast to the other approaches of virtuous friendships as a state of mind with selfless interest in the well-being of the other, these pragmatic criteria prescribe how such virtue friendships are observable, or how certain practical expectations of a virtue friendship can be justified: (1) mutuality of worldview, (2) authenticity, (3) equality of friendship relations, (4) diversity of interaction.

As Danaher also reconstructs the argument, these four elements are conditions for virtuous friendship. It seems obvious that machines can satisfy very few of these conditions: Machines do not possess a worldview, they cannot be "authentic", they are created, they do not age, perceptual categories such as "experience" and "memory" are fundamentally different in them, etc.

Danaher argues that points 3 and 4 are merely a technical challenge, i.e., we can create machines that meet these criteria. Points 1 and 2, on the other hand, seem to be metaphysical criteria that cannot be overcome by technological turn. The technical challenges of equality and diversity of interaction can be constructed by imposing certain limitations on the performance capability of a machine. For instance, the memory of a machine is modeled on the memory of a human being, e.g., by not repeating all possible old conversations on demand, but rather retrieving certain keywords or "memories" associatively.

However, especially points 1 and 2 are of manifold relevance: Danaher also admits that in the foreseeable future it will not be possible to create robots to which we can grant the mental states needed to be "authentic" and to have a certain "worldview" without reservations.

But how should virtuous friendships with machines be possible if they themselves cannot fulfill these requirements? Danaher's theory is based on the approach he calls "ethical behaviorism" (Danaher, 2019b). This is not concerned with the intentions of a moral actor, but merely with the behavior of those actors. Transferred to Aristotle's theory of virtuous character, this approach thus presents a pragmatic interpretation of virtues, which are inferred less about the mental states of actors. Rather, those virtuous character traits are inferred from their behavior.

In this way, Danaher argues, it is also possible for machines to enter into such a friendship: A consistent simulation of typical forms of worldview and authenticity, for example, by the individual, often idiosyncratic preferences and modes of communication, may suffice from a purely behavioral perspective to simulate the requirements of a virtuous character.

The idea that virtuous friendships can be reduced to the four criteria proposed by Danaher is controversial. However, the theory that moral action can be reduced to moral behavior is one escalation level higher: ethical behaviorism can be rejected on a broad front of arguments that start at a conceptual linkage of many ethical terms, such as responsibility to action, i.e., a purpose realization attempt, and end at practical questions of the utility of morality, such as the coordination of legitimate interests of some versus the necessary rules for everyone to get along.

One argument for not dismissing this idea out of hand is the justifiable assumption that these philosophical considerations will not necessarily catch on. For many people, a machine that is able to simulate ethical behavior is likely to be good enough to deal with these machines in such a way that they can be friends in a virtue friendship sense. It is also possible that the notion of authenticity can be extended via this pragmatic turn. A comparison is illustrative here, which serves as a kind of pragmatic test:

We assume that we are good friends with two persons. Their behavior is similar, they are open to our problems, helpful, friendly and also bring their own concerns and successes into our friendship. One day we learn that both friendships were not "real": One person was an actress who was paid to deliver a performance of a friend. The other person was not a person, but a machine in Danaher's sense, programmed to simulate a friendship.

The suggestion is that the "inauthenticity" of the machine is less problematic than that of the actress: the machine did not conceal any "actual" thoughts and feelings, since it has no thoughts or feelings except those it pretends to have. A machine cannot be "inauthentic" in this sense because it has no authentic alternative. Without question, the discovery that we have been friends with a machine will hurt. However, the question does not arise as to what the machine "really" thought of us the entire time. This problem appears differently in the case of the actress: even if we all play theater (cf. Goffman, 1959), it is possible to nevertheless productively speak of deception when we intentionally pretend something to someone else, such as a friendship. Here, then, the distinction between inauthenticity and authenticity is plausible.

A friendship with machines, in this sense, is not an in-authentic affair because there is no authentic "core" of the machine.

In an argument in the style of Descartes' skepticism of the external world, it can be stated that we also have no privileged access to the mental attitudes of our friends, but can only interpret their behavior, making the difference in the interpretation of "real friends" and "machine friends" a purely theoretical exercise. This phenomenological approach to friends, however, does not usually lead to paranoia about whether our friends are really interested in how we are doing, but is contextually embedded in such a way that we have no doubt that they are. It stands to reason that this may also be the case with machines, especially when one considers how children are already growing up with the idea that there is a non-human voice in the home that can control the lights and music. Insofar as there is a certain habitualization in which a machine asks us how we are doing every day after work, it can be assumed that many will not question this question at some point, but will answer it.

Thus, despite the philosophical reservations about ethical behaviorism, we can initially assume that virtuous friendships with machines can be theoretically simulated, and possibly even actually implemented in the longer term.

However, Sven Nyholm has already pointed out in his book Humans and Robots (Nyholm, 2020) that a mere metaphysical and technical perspective on the possibility of virtue friendships with machines is truncated: the ethical dimension is missing, which, according to Nyholm, must be understood as a similarly strict limitation as to the first two conditions (cf. ibid., 133). We will get back to these considerations later on.

5.7.2   Elder: The Most Excellent Friendship

Alexis Elder's work on "Friendship, Robots, and Social Media" (2018) and several of our theses argued for in this investigation are of similar nature and direction, especially in the question of whether the increased use of social media and thus the technological mediation of friendship will affect the quality of the latter (see Chapter 6).

Elder is mainly interested to find the conditions for the very most excellent friendships and whether we should be able to expect machines to ever reach this level. The clear answer for this is "no". For this answer, she presents a two-step argument that is leading to the conclusion that machines may not be candidates for the most excellent friendships. First, she determines that virtue is necessarily connected to the very best possible friendship. And second, she argues that the conditions to participate in virtue must remain out of reach for machines in this meaningful sense as this would amount to false coinage.

For the first step, she puts forward a rather simple point: friends usually are not only interested in closeness and intimacy but also in the other person's well-being. If friends are, however, interested in the well-being of a person, then they must be interested in the factual improvement of someone's life rather than merely pleasing them. They must have an idea of what is good, and thus must have an idea of what is virtuous. In her view, then, "bad people cannot be good friends" (ibid., 56–67) because bad people are not interested in the actual improvement of the friend's life, even though they might be interested in the other person's temporary benefit. Bad people can even be close friends or best friends in the sense of knowing the other person intimately and caring about their subjective thriving. However, as bad people do not care about the good, neither for themselves nor for others (otherwise they would not be bad people), they are limited in pleasing the other person rather than elevating their well-being (though this might coincide from time to time).

Her approach to motivate this first insight is thus a test of the negative: in order to find the conditions for friendship, she tests them against people who are otherwise immoral. She grants that morality must not be the only measure of the quality of friendship (ibid., 59). An overly moralized friendship will not be considered especially good as we should seek closeness as well. Neither from the phenomenology of what we consider especially excellent friendships nor from the consequences of widespread paternalism in these relationships. If our concern was primarily with other people's well-being as a marker of friendship, we would have to constantly disagree with many of their decisions and aim to undermine their decision by correcting them for what we deem the most conducive to their well-being. Thus, Elder refers to the generally unstructured form of friendships, in which acknowledging the other person's subjective preferences is part of valuing the friend. Even if we consider someone's preferences or behaviors "vicious" we can remain good friends with them, as long as their vices do not affect the friendship or us. However, as she points out, valuing virtue must be part of the friendship: generally, more virtuous people will lead better friendships, and generally more vicious ones will lead worse friendships.

She later seems to commit to the rather unfortunate consequence (in agreement with Aristotle, 2000, 1156b; Brink, 1999) that in her view "the most excellent friendships" must remain unattainable to many humans, as the requirement for a dedicated, intimate interest in the other person's well-being ought to be paramount as a reflection of a virtuous character (cf. Elder, 2018, 67). This standard of character seems to be rather rare.

In the second step, these necessary abilities to participate in virtue are being tested on machines. Elder takes Aristotle's argument of comparing fake friends to fake coinage: If we merely want to make a friend feel better about their financial situation, giving them counterfeit money would be an appropriate thing to do. Without actually having to share our wealth, or do anything of particular effort, we could "improve" their situation by providing them with pretend-money without them knowing. They will feel better, and potentially never find out if their main goal is to feel relieved of some financial worry (ibid., 84). However, as they are not really rich, their situation has not been improved at all besides being blindsided.

Elder claims that machines will not be able to provide anything more than what counterfeit money would do for the case of the financially worried person. This is both because, from the first point, machines will not be genuinely, deeply interested in the person's welfare as they do not have any reason to be: as they have themselves no sense of well-being, they have no reason to respond to this condition of friendship. Further, even if we programmed machines to care, we would provide those who seek their attention merely with such fake attention. The machine, lacking virtue, could pretend to be interested in the person's well-being by catering to their needs, but any more serious understanding of what a person would need for their well-being in the virtuous sense, i.e., against their subjective wishes and with a certain paternalistic self-confidence must fail. A machine does not "know" what actual well-being would amount to, as the machine is lacking virtue. While Elder grants that even fake attention can be helpful, and some human–human interactions could be replaced (ibid., 86), the general thrust of the argument, and ultimately the rejection of human–machine friendships, lies in the fact that the attention and value one receives from human–machine friendships is deceptive.

With Elder's approach to virtue friendship and their ability (or rather inability) to be transferred to machines, we have one of the most elaborate approaches to virtue machine friendships at hand. Next to Danaher's reduction of virtue friendships to specific requirements and arguing in favor of machines entering these relationships, Elder's position is the opposite: the fundamental lack of understanding of what it means to seek someone else's well-being due to a lack of moral character must lead to machines having to remain outside this friendship circles. Most excellent friendships, thus, are rare and only possible between humans.

There are at least two problems with this argument. First, it remains rather unclear how the negotiation of the different requirements, arguably based on the "standards of character" (ibid., 68) will lead to the reliable outcome that will mark some friendships as most excellent and others as less so. There are, usually, many more values to be fulfilled or instantiated in these friendships. Elder is merely discussing what is a necessary condition to be able to talk about virtue friendships; we should, for such a far-reaching judgment of the "most excellent" forms of friendship, expect that she would give us sufficient conditions for such judgment, unless we want to be pluralists about these "highest forms" of friendship. If, however, we are ultimately pluralists about the most excellent friendships, it seems rather odd to declare many friendships less than excellent.

This argument is also taken up by Danaher's own proposal of how machines can indeed fulfill virtue-friendship conditions as he breaks down the presented arguments for virtue friendships into four necessary conditions. Once these are fulfilled, a virtue friendship has been established. And while we can see that a committed concern for the other person's well-being is an important part of such a friendship (and a difficult task for machines to complete), it cannot be the only requirement. Otherwise, we would merely create a (rather large) class of friendships as "best" ones. As Elder admits, however, most excellent friendships are rather rare to begin with. This brings us to our second concern.

In this second point, we should take a closer look at the admission that "the best friendships" will be limited to the most virtuous people, and thus to the moral elite. Following logically from the arguments from virtue theory, only two people who are both of high virtue can enter and maintain those excellent friendships, as the mutuality and equality of such friendship are both presupposing virtue from both parties.

However, this must also mean two things: first, we can declare two people in very close relationships that are not always fully committed to each other's well-being to be non-excellent friends even though both parties may be very happy in them nonetheless. The necessary commitment to constantly care for each other's well-being might not always be present in some friendships, even though both parties would still consider them best friends. I believe Elder will endorse this as not a bug but a feature of virtue theories of friendship: we should be able to tell people if they are in a virtue friendship. If you do not primarily care for each other's well-being in the eudaemonia sense, then your friendship is not the best one possible. But what does this kind of assessment help us in judging other people's friendships and nudging them to lead better lives? It seems odd to make positive suggestions to people on how to structure their friendships (even though, as pointed out before, Elder merely provides a necessary condition rather than actually a list of what makes it excellent, as this seems to be supposed to merely follow from virtue). Especially if people have other connections, such as family relationships, that care deeply about each other, they might not require the most excellent friendships to be part of their lives.

And second, a concern coming from the negation problem: Elder seems to suggest that faultless ends to friendships are impossible. There is a variety of standards of character and of understanding of each other's well-being, and especially in understanding of what it means to be vicious. We can thus imagine two friends growing apart by changing their fundamental understanding of what it means to be good, and while they remain committed to each other's well-being. This understanding of each other's well-being, however, can conflict to a degree that makes a friendship impossible, without the presented theory being able to tell who's at fault. As we have seen above, Elder even suggests that friendships should end when one person is vicious and does not want to change their mind. It seems that if two friends have changed their minds for the better, becoming both more virtuous, this could mean the end of their friendship as well. A consequence she does not seem to be able to explain.

5.7.3   Ryland's Twist: Degrees-of-Friendship?

In opposition to Elder's approach stands Ryland's work on friendship. Her theory (2022) suggests that friendships generally should be considered to come in degrees, rather than in kinds, and that from a certain sufficiency requirement we can consider friendships established. However, when she talks about friendship, she typically means merely virtue friendships. Her pre-stated goal is to propose an understanding of friendship in which all Aristotelian types of friendship—utility, pleasure, and virtue—are based on the same kind of principle of minimum requirements. Thus, the three types are not to be understood as adding onto each other until we achieve the highest level and kind of friendship (in forms of a certain virtue friendship), but we can have different levels of each of these kinds of friendships to any kind of minimal and maximal degree. This way, she offers this kind of pluralist approach that we required from Elder.

For this, she enumerates and consequently rejects all the previously proposed necessary conditions for virtue friendship. She mentions reciprocity, empathy, self-knowledge, shared activities, associative duties, affection and well-wishing, love and admiration, based on duty, honesty, and equality. Every single condition is being rejected on the grounds that some of them have obviously fallen out of time, that some others are clearly not necessary, and that some of them can be interpreted in a more performative-pragmatic way without incurring too many ontological concerns.

Thus, she rejects the often considered "necessary" conditions of Aristotelian virtue friendships by pointing toward none of them resonating with what we would call a friendship. She picks up some of the themes in other conceptions of friendship that go without the Aristotelian picture. As her declared goal is the lowering of friendship standards to include machines, these standards should not stand in the way based on some humanist or speciesist assumptions about friendship. Her proposal comes twofold: first, she is understanding all the previously necessary conditions to be merely sufficient conditions for specific kinds of friendships, as these conditions do not decide about the fact whether or not there is a friendship, but rather about the specific quality of the friendships. Honesty, shared activity, or some other ingredient might appear necessary to us to consider a friendship between two people "virtuous", but ultimately, we can always imagine different kinds of friendship participating in minimal requirements of friendship even though they may not exhibit these conditions. Her proposal to capture these minimum requirements consists of the concern for "mutual goodwill". Mutual goodwill is considered to do all the work necessary conditions for friendship do without having to commit to anything more ontologically or psychologically demanding:

As she also elaborates, we can even consider these kinds of friendships established in the current technosphere, as we can assume that the machine-part of her "mutual goodwill"-condition will be fulfilled by a social machine not intending to harm us (Ryland, 2022, FN 23). Thus, some people can claim at this stage to be friends with machines, as they fulfill the necessary conditions for beings friends with each other, while also, additionally, fulfill some sufficient conditions in form of other demands for a friendship. Even though we might be friends with machines on the lowest of degrees, we can still be friends with them in these pictures, while Aristotle and others would have rejected the idea of these friendships outright.

However, while Ryland has developed similar ideas to those that will be presented here, especially in rejecting the standard conditions for friendships in the Aristotelian sense, we should confront her approach with some criticism as well. This criticism leads us to the "synthetic friendship" approach in this chapter, as we develop some negative conditions and positive suggestions (akin to Rylands) to read into human–machine friendships. However, ours will be an approach of different kinds, not different degrees.

She explicitly states that she is after "degrees-of-friendship" over the idea of "kinds of friendship", as she advocates for the ability for being friends with machines on a rather low degree, but a degree nonetheless. However, as the twist for her proposal on how these degrees are minimally achieved (i.e., how to distinguish a friend from anyone), she proposes that we should show goodwill to each other. This good-will is intended to replace the other conditions fully while being open for other sufficient conditions down the line.

We should ask two questions to such an account in order to understand it better without having to reject it immediately. First, we should wonder what is achieved if we fulfill this first condition, and what follows from this fulfillment. And second, whether we can reasonably assume that machines will have a "goodwill" toward us that is not sneaking other strong conditions about machines, and thus about friendships, into the debate. The first argument concerns the fact that we can be mutually good-willed toward anyone. Imagine the supermarket cashier that you see once a week, or your neighbor that you greet friendly, or even your coworker that you see online on a regular basis: for all three you have reasons to hope for their well-being. Partially because their well-being will benefit you, but also as a good person it is not unreasonable to assume that these people's well-being is of genuine concern for you. If we assume that the other ones are equally caring, we should assume that you and these people are fulfilled the one necessary condition for friendship. Now imagine that you have a weird thing about not lying to people you meet on the street (for whatever reason), and the supermarket cashier half-interestedly inquires about your state of affairs. Your honest answer might inspire an honest response, and the two of you have a sincere exchange about life. You both nod, and go about each other's life, never to return to that moment, reverting back to the mutually goodwilled nod and appreciation. Are you friends? In a close reading of Ryland, she would have to acknowledge that you indeed are. At least to such a small degree, potentially mere trace-amounts, that some or another sufficiency condition for friendship has been fulfilled. Thus, in this close reading, we should consider this to have been a "friendship". We will see down the line that these vapid, mere occurrences of friendship are not a good predictor of friendship, yet Ryland is committed to call these to be of the "lowest degree" of friendship.

The second point explores Ryland's somewhat self-evident assumption about the ability of machines to exhibit goodwill. While we generally may know what is meant with "mutually goodwill", we can see that attempting to nail down this particular condition will affect the strength of Ryland's ultimate thesis, i.e., that we will be (and potentially are) capable of having friendships with machines. To read her as charitable as possible, we should consider two possible readings of what a "mutual goodwill" could mean.

The first reading involves a similar approach to Danaher's ethical behaviorism (Danaher, 2019b). In this idea, the very fact that something behaves a certain way consistently and reliably should provide a reason that this entity is acting ethically. This shows mostly in our ability to determine whether a machine wants us some harm or if they avoid doing so. If a machine is capable of measuring its impact on humans and thus stop or refrain from doing something that would most likely harm us, there is an intuitive appeal about the machine not wanting to harm us. Ryland's explanation for her idea of mutual goodwill appears to go into that specific direction, as she rejects a potential concern for "ill-willed machines" to be based on a rather superficial understanding of contemporary robotic development. A machine that is programmed to not want us harm, as most machines do not, thus, is a machine with good mutual will.

However, even in such a problematic reading (for other purposes, i.e., how would we even know about the behavioral background of such actions), we may encounter some even more problematic consequences. Ryland herself claims (Ryland, 2022, Footnote 23) that an ill-will is "particularly unlikely" to be present in the kinds of robots we are currently interaction with. We should wonder how we can find out about the likelihood, and thus unsupervised emergence, of such an ill-will. If we are merely dealing with a behavioral feature of a machines which is primarily devised by human interference, then we should not expect for such an apparent ill-will to emerge in the first place. Because if a machine starts becoming "unfriendly" toward us, we can merely point toward such unfriendliness and demand a correction of it. Thus, the mutual goodwill cannot be more than a technological pre-configuration without any connection the standard conception of what makes a goodwill.

The second reading rejects this behaviorism outright with the argument from Ryland (2022) that a psychopath, who is merely intending to use us in their long-term plan, pretends to be our friend without us being able to tell the difference. Elder and Ryland agree that this person could never count as a friend, as this person conceptualized us as a victim of their purposes, and Ryland considers "considering someone as a victim" and "considering someone as a friend" are conceptually conflicting. It is, in this regard, necessary that "mutual goodwill" is constituted by each party only intending the best consequences for each other. However, we may ask how specifically this "intending for the best consequences" can be measured, as we have a hard time with most technologies to read any kind of intention into them. Would a Roomba-cleaning robot count as "good-willed" if they avoid running over my foot while cleaning? Ryland appears to say so. Whether or not a friendship emerges from that is up for debate, as the sufficiency requirements are coming into play—yet, we are left with the idea that a robot may be good-willed toward us.

Thus, concluding Ryland's approach about the degrees-of-friendship view, we may take much of her work to be a useful contribution for the question of whether machines and humans can be friends. In more than a trivial sense it appears that we can, and Ryland attempted to make light of an otherwise unclear situation from what follows after this trivial insight. The idea that we can lift the difference between human–human friendship and human–machine friendship *in kind* by negotiating both down to a necessary condition marking a difference *in degree* seems to be throwing out the baby with the bathwater.

5.8   CONCLUSION

Some friendships are better than others. That is a philosophical problem because we should find out which ones are better and which ones are worse, and why. This has motivated different theories about the value of and within friendships and has led to different schools of thought on how to lead friendships. The dominating premise of friendship-theories is also the first one to have ever been proposed at length: the one in which friendships and virtue are necessarily connected, in which friendship quality is a consequence of the quality of character of the friends. We have seen that there is a problem when trying to understand what it means to end such a friendship, as virtue would not suggest giving up or ending friendships, even though there are situations in which faultless ends of friendships are considerable.

This virtue paradigm has led modern researchers to apply this theory to the question of human–machine friendships as well, with a variety of outcomes. While Danaher confirms the possibility of such friendships,

Elder denies them. Ryland, in the end, reconsiders the basis of what a virtue friendship requires to show that we can be in (growing) degrees-of-friendships with machines. As none of them have been without their issues, we might have reason to seek an alternative.

REFERENCES

Allan, G. (1998). Friendship, sociology and social structure. *Journal of Social and Personal Relationships, 15*(5), 685–702. https://doi.org/10.1177/026 5407598155007

Aristotle. (2000). *Nicomachean ethics* (R. Crisp, Ed. and Trans.). Cambridge: Cambridge University Press.

Brink, D. O. (1999). Eudaimonism, love and friendship, and political community. *Social Philosophy and Policy, 16*(1), 252–289. https://doi.org/10.1017/S02 65052500002323

Coates, D. J., & Swenson, P. (2013). Reasons-responsiveness and degrees of responsibility. *Philosophical Studies, 165,* 629–645. https://doi.org/10.1007/s11098-012-9969-5

Danaher, J. (2019a). The philosophical case for human-machine friendship. *Journal of Posthuman Studies, 3.*

Danaher, J. (2019b). Welcoming robots into the moral circle: A defence of ethical behaviorism. *Science and Engineering Ethics, 26*(4), 2023–2049.

de Graaf, M. M. A. (2016). An ethical evaluation of human–robot relationships International. *Journal of Social Robotics, 8,* 589–598. https://doi.org/10.1007/s12369-016-0368-5

Elder, A. (2014). Excellent online friendships: An Aristotelian defense of social media. *Ethics and Information Technology, 16*(4), 287–297.

Elder, A. (2018). *Friendships, robots, and social media.* Routledge.

Goffman, E. (1959). *The presentation of self in everyday life.* Double Day.

Hall, J. A. (2012). Friendship standards: The dimensions of ideal expectations. *Journal of Social and Personal Relationships, 29,* 884–907.

Helm, B. (2022). Friendship. Stanford encyclopedia of philosophy. https://plato.stanford.edu/entries/friendship/. Last accessed 15 June 2022.

Levin, N. J., Kattari, S. K., Piellusch, E. K., & Watson, E. (2020). "We just take care of each other": Navigating "chosen family" in the context of health, illness, and the mutual provision of care amongst queer and transgender young adults. *International Journal of Environmental Research and Public Health, 17*(19), 7346.

Lewis, C. S. (1988). *The four loves.* Harcourt Brace.

McCabe, J. (2016). Friends with academic benefits. *Contexts, 15*(3), 22–29. https://doi.org/10.1177/1536504216662237

Nehamas, A. (2010). The good of friendship. In *Proceedings of the Aristotelian Society*. New Series (Vol. 110, pp. 267–294). Blackwell Publishing.

Nehamas, A. (2016). *On friendship*. Basic Books.

Nyholm, S. (2020). *Humans and robots: Ethics, agency, and anthropomorphism*. Rowman and Littlefield.

Ryland, H. (2022). It's friendship, Jim, but not as we know it: A degrees-of-friendship view of human–robot friendships. *Minds and Machines, 31*(3), 377–393. https://doi.org/10.1007/s11023-021-09560-z

Sherman, N. (1987). Aristotle on friendship and the shared life. *Philosophy and Phenomenological Research, 47*, 589–613.

Smith, J. K. (2021). *Robotic persons*. WestBow.

Thomas, L. (2013). The character of friendship. In D. Caluori (Ed.), *Thinking about friendship: Historical and contemporary perspectives* (pp. 30–46). Palgrave Macmillan.

Vela-McConnell, J. (2017). The sociology of friendship. In K. Korgen (Ed.), *The Cambridge handbook of sociology: Specialty and interdisciplinary studies* (pp. 229–236). Cambridge University Press. https://doi.org/10.1017/978 1316418369.024

Wittgenstein, L. (1953). *Philosophische Untersuchungen*. Suhrkamp.

CHAPTER 6

Digital Hermits

This Chapter analyzes an emerging lifestyle, the one of "digital hermits". We understand those digital hermits to prefer seclusion from their immediate physical surroundings over conventional socializing, and to seek human connection through digitally mediated connections. In analyzing this lifestyle, we find that digital hermits do not have to be pathologized as sociopaths, but that their atypical social behavior may become another standard for human relationships to flourish. We might even see improvements in conversations that are both instant and ubiquitous, yet text-based and filled with exchanges about each other's thoughts and emotions. We conclude that digital-only friendships are likely to grow, philosophically unproblematic, and ultimately merely the acknowledgment of social connections evolving through technological progress.

Humanity has been through a lot of changes that seem to be somewhat independent of each other but may be connected in a more long-term view of our shared future. One of those changes is the increased prowess of AI-based applications, especially language models and their uses. While many of the big companies remain rather hermetic about peer-reviewable benchmarks and constructions of their larger language models, we will see an increase in language-based tools utilizing large language models (or smaller ones that are similarly effective in specific subsets of tasks).

Another change was and is the pandemic of Covid-19, which has not only cost millions of lives worldwide (and counting) but also has,

H. Kempt, *Synthetic Friends*, https://doi.org/10.1007/978-3-031-13631-3_6

arguably, put more than just 1.5m distance between many people, in all kinds of social contexts: from heartbreaking stories of lonely birthdays, canceled graduation parties, delayed marriages, collapsing tourism and any kind of entertainment and creative activity, weeks and months of self-isolation for physical risk-groups, distance- and home-teaching fueling social class differences. The world has become a more distant place.

This enumeration demonstrates, however, that abrupt and harmful changes in our modes of life are not upending lifestyles, but rather they are demanding adjustments, often accelerating changes that have been merely lying dormant before. And one demonstration of such change in the pandemic has been the ability to connect digitally. Not as people's first choice, but as a necessity to maintain social contacts of all different kinds.

For some, the ability and to some degree necessity of connecting with others primarily digitally has been part of their social life for a while. For now, we can call those who conduct most of their social connections online digital hermits. Digital hermits require very little physical-presentist social interaction, while not being overall less sociable than others: in fact, they might be even more sociable in their own sense of sociality—open for others, for their lives, opinions, and interests. We also do not have to assume that digital hermits do not leave their house (though some indeed never or rarely do)—they are not asocial in the sense that they avoid humanity (which would make them an actual hermit), but they are asocial in the sense that the immediateness of human interactions, especially more cognitively demanding, emotionally intimate, and socially complex ones may not suit them well.

In this Chapter, we will explore first the nature of being a hermit. After that, we will transfer the concept of hermit to the digital age and explore whether we can make sense of the notion of "digital hermit". Lastly, we will discuss whether people who choose to be digital hermit are missing out on life and if it affects their ability to be friends with others. In this, we will be supported by Alexis Elder's investigation on whether technologically mediated friendships could still be considered virtue friends. Additionally, the previously discussed approach by Helen Ryland into degrees-of-friendships will factor in our assessment as well, even if we ultimately reject the idea of there only being degrees-of-friendships (rather than kinds).

6.1 A Word on Hermits

Hermitdom has long been a part of both philosophical and religious practices. In pursuit of wisdom, enlightenment, and other purposes, retracting from society either temporarily or fully has been a way to achieve these purposes, and often been required as a lifestyle to maintain these levels of wisdom. Hermits have been recognized as sages and have been excused in their lifestyles as they rarely contribute to society besides the occasional teaching of their insights borne from seclusion. Thus, the key point about hermits is the rejection of certain levels of sociality over a presumed superiority of being alone. The phenomenon of hermits is useful as a measure of the levels of sociality, as a positive definition appears more difficult to come by: if we want to characterize what sociality includes, we might have to list too many different forms of connections as examples of sociality, especially when considering that our sociality is usually considered constitutive for lots of other elements of personhood (see Chapter 4). If we, however, look at those who reject social connections, we might be able to gain some insights into the conditions of certain social contexts. In this way, we should first investigate what is to be lost in sociality when one rejects society.

6.2 Ways of Being a Hermit

To gain a better understanding of what it means to be a hermit, we should take a closer look at some of the different ways of being a hermit. This is not an analytical attempt in carving out precisely what it is to be a hermit, but rather a collection of behaviors and preferences that are generally counted as "asocial", "introvert", or some other non-socially-integrative behavior. From a terminological setup it appears useful, then, to have an umbrella term designated to differentiate kinds of "asocial" or "unsocial" behavior without pathologizing or evaluating these behaviors (see for example the phenomenon of "hikikomori" as "acute social withdrawal" ([Bowker et al., 2019; Hamasaki et al., 2021]). The term "hermit" appears useful for this purpose. Generally speaking, what we mean with hermit is someone who removes themselves from some or all social connections for meditative, social, psychological, moral, religious, or other reasons. We can identify four features of such removal from social connections to clarify the different ways of being a hermit.

We can, first, imagine different hermits with different strengths of rejection of social relations. The classic type of hermit is usually a religious, reclusive person seeking wisdom or insight from removing themselves from society. In this sense, becoming a hermit is a meditative praxis. Nietzsche's Zarathustra is an instructive example for this kind of hermit, as Nietzsche's character fully removes himself from society to only teach, upon returning to society, his insights gathered in his exile. The feature here is the sought-out distance to other people and thus the lack of possibility or opportunity to encounter others. However, other people we could also reasonably call hermits who are not necessarily removing themselves from society by removing the chances to interact. They merely do not actively seek connection to others and thus, however slowly, drift apart from society.

Thereby, it seems fair to characterize hermitdom as a spectrum of rejecting sociality: the strongest hermits are not only those not seeking social contacts but are actively rejecting social connections in retracting into the remote wilderness, while less strong flavors merely drift apart from society. We can call this the intentionality of hermitdom, in which the attempts in avoiding social connections play a role in their lack of sociality. The more committed and intentional a hermit is in seeking solitude, the more likely will they succeed in the long term.

There is, second, the dimension of quality of connection. One can reject human connections but welcome, for example, ones with dogs or other animals, or reject certain kinds of human connections in favor of others. Diogenes might be a good example here, as he did not reject coming into contact with other people (thus counting as a "weaker" version of the first feature), but preferred not to establish any kind of deeper connection. Thus, one can be a "hermit among people", so to speak, if someone chooses not to engage further with their surroundings, while not being isolated. One can care deeply for animals and nature, and one can even admit on depending on relating to their human society to some degree.

We may want to call this feature the quality of hermitdom, as we should account for those who acknowledge the necessity of society but prefer to "keep it light" with other people.

Third, next to the intentionality and quality of rejecting connections there is quantity. If I only aim to have social connections with two people while rejecting getting in contact with anyone else, I would still be considered a hermit in this quantitative sense. This feature of hermits may be more often described with introvert personalities rather than hermitdom; however it can compound to not seeking relationships with people, or one may in fact reject society as a whole but is committed to stay in relationships with a small number of people. Think of the image of an artist seeking solitude while remaining happily married or a parent, or of someone uninterested in getting to know anyone new. The rejection of wishing to get to know new people could be, in this light, understood as a feature of quantitative hermits (recall that hermit is merely an umbrella term for a variation of anti-social behavior).

Lastly, we might want to assign a certain temporality to hermits. Usually, hermits are considered long-term rejectors of sociality, as the removal from social connections is a fundamental change in lifestyle that requires preparation, life adjustments, and a certain commitment of the purpose to follow through. However, we can also see that a rejection of sociality can be a mere temporary occurrence in which people simply do not see the urge or need to meet new people for an extended period of time and for different reasons (grief, self-care, general re-orientation in life). One could argue that too short a time of these anti-social wishes should not count as hermitdom. This is true, as an evening alone should not amount to being a hermit. However, it seems that the choice of staying alone for a full weekend instead of interacting with anyone is telling us something about the sociality of that person.

Thus, when describing the social connections of a person (or the lack thereof), we may want to consider the intentionality, quality, quantity, and temporality of these connections in order to assess whether it is deserving of the term hermit. In the following, we will use this term to characterize different kinds of people and their social connections to then test whether the implicit thesis, that having quality friendships and being hermits are somewhat conflicting aims, holds true. At this stage it is noteworthy that Alexis Elder argues for a similar goal (Elder, 2014, 2018, p. 139): in her argument, she tests whether technologically mediated friendships can be excellent friendships in the Aristotelian sense. Thus, we should feel in good company to investigate whether friendships through machines can be any less than analog friendships.

6.3   What Hermits Are not

Our concept of hermit has thus far been rather quiet on the specific motivational reasons for rejection sociality. While we acknowledged a few and ruled out the necessity for another (e.g., mental illness), I argue that we should pursue an inclusive or liberal concept of hermitdom. We thus can rule our some otherwise "necessary" conditions as, at best, sufficient reasons for calling someone a hermit.

The main one here would be that seeking solitude is somewhat a statement on the person's view of society. While we can assume that most hermits do not view society too favorably, we do not have to assume that every hermit is a misanthropist. The idea that society has negative effects on those seeking enlightenment is a pervasive story in religious hermits (and their solitude is intended to purify their soul), but certainly not a necessary condition for leaving society. I can be fully committed to working on my enlightenment, too, without having to assume that society is otherwise ruining my path (as well as I can assume that society is rotten and yet remain a part of it). Other reasons for (temporarily) leaving society can be obligations or distractions, for example. Thus, hermits are not necessarily misanthropic, frustrated, or otherwise negatively affected by society necessitating the avoidance of others to remain or return to being good. The truth is that the rejection of sociality can have many reasons and none of them should count as necessary to be considered a hermit. If someone willfully rejects sociality in one form or another, we should consider them a hermit.

6.4   A Fifth Feature of Being a Hermit: Digital Only

After having elaborated on the different kinds of hermits, we can move toward their digital versions. This is relevant to the debate surrounding friendships as we can, in the step after that, assess the ability of a digital hermit to be a friend to others. From the insights to that we may be able to transfer some conditions or suggestions about purely digital friendships between two hermits onto friendships between a human and a machine.

We previously distinguished four different types of features of being a hermit: we can intentionally reject sociality, we can reject certain qualities of sociality, certain quantities, and for a certain amount of time. We can now consider a fifth feature, which has not only seen an externally motivated growth due to the pandemic of the early 2020s but also is being carried by the growing technological options to stay in contact with other people. We may call this the "digital hermit": a person that prefers to not have physical contact with people, if at all, but is fine with using digital technologies to have mediated contacts with others.

This feature lies somewhat at odds with the other ones mentioned so far, and we should clarify some potential misunderstandings.

First, digital hermitdom is not intended to introduce a full new layer of hermitdom but rather should be seen as an additional feature: someone barely using any social media is not an "analog hermit" due to their refusal to engage digitally with people. Rather, it shows that preferring digital sociality over the analog one is still a norm violation in terms of anti-social behavior. Previous pathologies of "internet-addiction" demonstrated this mistaken approach to sociality (Kempt, 2020, p. 29) and have shown that the standard expectation of pro-social behavior is in-person sociality. This discourse has been occurring in every decade thus far in which new social media allowed for (re-)creations of avatar-based social interactions. From chatrooms and instant messengers in the 1990s and 2000s to "Second Life" and other MMORPGs and social media platforms in the 2010s to the current debate on the "metaverse" (Chalmers, 2022), the struggle surrounding the recognition of preferring technologically mediated social connections has been controversial. The intention to name this phenomenon is the reason to use the term "digital hermits".

Second, as the other features combine (one can be a hermit both in the qualitative and the temporary sense), so does the digital dimension. However, this only applies one way, as stated under the previous point. One can be a weak intentional, digital hermit who prefers to have most of his social connections to be digital, but is not actively avoiding in-person social connections. However, as we stated before, it would be weird to call anyone who does not use social media but otherwise lives a standard, socially integrated life to be a "analog hermit". The norm suggests that this is not a second layer of sociality yet. Though, this may eventually change in which case "digital hermits" will become indeed a second layer of sociality which interacts with the first layer almost equally in terms of norms of participation.

Third, one may claim that digital hermitdom is merely a subcategory of the quality feature of hermits, as we have quality implications for preferring purely digital connections over analog ones. It could thus merely be described as "choosing not to have all possible quality connections".

However, this denies the fact that one can have essentially all kinds of relationships with other people without ever meeting them in person. Thus, I can also be a digital hermit and one in the qualitative sense without doubling the semantics: I can reject in-person socialization but additionally also reject the idea of ever interacting with someone with small talk. This might make me rather difficult a person to interact with, but hermits are generally a rather difficult group of people. In fact, the question whether digital hermits are lacking in the quality of their social connections—willingly or unwillingly—is one of the key issues in considering friendships. This will be discussed in the following.

6.5   Hermits and Friendship

Why talk about analog and digital hermits at all? The reasoning behind distinguishing between analog and digital hermits can be made clear by looking at how people exhibiting these traits are judged. Usually, for most accounts, we judge those who prefer to stay socially aloof or uncommitted, who disappear for some time to seek solitude, or who straight-up reject social connections to be lacking some of the necessary features to be good friends, if they are considered friend-material at all. With some of the theories presented before, it appears that many philosophers would judge digital hermits, even if the preference of digital over in-person social connections is the only feature of their hermitdom, to possess a smaller potential to be a good friend than those present in person.

Because as far as we have seen, many established philosophical theories about friendship judge hermits harshly. Effectively, being unwilling to meet people in person, so seems to be the tenor of several contemporary theories of friendship, makes one a less good friend (see, for example, Cocking & Matthews, 2000, Fröding & Peterson, 2012; McFall, 2012). Elder and Ryland show impressively that this assessment does not affect the way of being a virtue friend on principle, but rather that the most excellent friendship simply is not possible with these conditions, while Kaliarnta shows, similar to our approach, how the account of virtue friendship can actually be misleading in the assessment of online friendships (Kaliarnta, 2016).

Elder is the most explicit in her assessment of the ability of digital hermits to be capable of leading excellent friendships. Her point goes to what Aristotle claimed the moral character is the primary concern for virtue friends and the interest in the other person's well-being. As such the fact that a person, and in fact the friendship, can be agreeing on the conditions of the interactions remaining fully digital (as they are predicated on equal recognition of each other's needs and wishes), can justify the rejection of the in-person criterion for virtue friends. While analog friends can be more present and thus better, the fact that someone may not be present does not affect their ability to contribute to a very good friendship.

Ryland, as we have seen before, recreates friendships as a question of degrees, and as long as "mutual goodwill" is present, every other feature merely adds to the quality of the friendship. It is clear, in her account, how a person rejecting some relevant activities of human interaction (i.e., physical presence) might limit the maximum degree of friendship they can enjoy with someone else. And even if we can question whether that is correct (we might even think that there are even better reasons to aim for a purely technologically mediated friendship), the fact that their friendship can still provide everything both friends can wish for should mean that it can be a friendship of high degree.

Even Danaher could have an account for digital hermit being good friends, though indirectly as a consequence of this consideration of whether machines could fulfill the necessary conditions for virtue friendships. His re-interpretation of the conditions to make them fit a machine's capacities includes the ability for a digital hermit to become a virtue friend as well. The idea that the "variation of interactive instances" (Danaher, 2019, Sherman, 1987, even more rigorous McFall, 2012) could be interpreted as a more realistic one, in which virtue friends are not constantly partaking in our activities, but rather that the quality of those activities is more important to us. We can imagine two people having quality times together, in which they collaborate on an art project, or discuss a movie they just watched together, or where they help each other to buy clothes they like. All this can be done without having to ever meet in person and thus fulfill the hermit-condition of this argument. For Danaher, this variation and availability might be sufficient to consider this a virtue friendship.

6.5.1   Four Arguments in Favor of Digital Hermits

Next to these considerations, we can add a few more that strengthen the case for purely digital friendships between, for example, two digital hermits. For this we will consider first the necessity for conversation over other activities, second, that some circumstances can force us to be digital hermits, and third that purely digital friendships can allow for a much wider range of personalities, thus increasing the potential for a more reflected, worldly, and experienced character, and lastly that some groups are especially benefitting such cultural shift.

First, considering what is often intended as a derogatory comment about people merely texting with each other all day instead of seeing each other in the eye. This judgment implies that a personal meeting with someone adds so much more to a friendship than a string of text messages ever could (see, e.g., Sharp, 2012). In many instances, we may agree. However, considering the practical pressures and necessities of purely digital connections may change our judgment about some of these circumstances. If a friendship (or a romantic relationship, for that matter) consists mostly of text- and vocal exchanges, these exchanges should be filled with content. While a shared game of tennis or a night at a bar has conversational dimensions, for the digital friendships the standard is often reversed: the main connection between two people who never see each other in person is their conversations. This not only enables but almost forces those friends to talk more, increasing the chances of talking more about each other, one's thoughts, feelings, and general dispositions to life.

Of course, these observations are neither a logical consequence of being in a technologically mediated friendship nor might these be the standards of those friendships. They also do not suggest that just because two people have nothing but the conversation then they will use that medium to say anything that would bring them closer together. However, this argument is intended to show that there are ways in which technologically mediated friendships can lead to an even more intimate, honest, and thus deep connection between two people. The argument that presence is something inherently more intimate should be rejected. A conversation, ongoing for years, can lead to an even more intimate connection with someone.

Second, the Covid-19 pandemic has impressively shown that our established habits of leading friendships might be less safe and more volatile than we previously thought. The fact that for several months in 2020 (and partially even 2021 and 2022), in many regions of the world, safety measures required a highly reduced contact with many friends of ours necessitated a pivot toward digital formats of hanging out with each other. For some people, this has been their lived reality before, and for some more, it has remained their lived reality ever since. We would not consider society to have become less one of friendships because our connections were largely shifted toward the digital sphere.

One might interject that a lot of people have shown a certain zoom-fatigue (Bailenson, 2021), as they require human contact and presence, and that this has been an extraordinary moment requiring everyone's willingness to reinvent their relationships (including families and friends). Thus, this would tell us little about the ability to have purely digital relationships.

This seems likely true even beyond the rather anecdotal evidence we have for it, as the participation in any kind of screen-based activity with one's camera turned on has a draining effect on the participants. This may then lead to most people rejecting this form of communication and relationship-building as they find it exhausting, limiting, and overly complicated. They would prefer to simply meet for a drink, thus the cost–benefit analysis would suggest that most people will not lead relationships like that.

This calculus does not extend to those who have no other option, however. Immunocompromised people have lived several years with these constraints, and since the world has become a more dangerous place for those who cannot get vaccinated against these diseases but will likely have strong complications if they become sick, they cannot calculate in these cost–benefit terms. Thus, the potential side effects of having to conduct friendships online will be contextualized to a positive overall preference. It would be odd to argue that those who are committed to our well-being and our friendship but are limited by physical constraints on their lives have to be in less good friendships.

Third, we may want to consider the biggest life-improving aspect of being able to connect with people purely digitally without the necessity for personal presence: any person with this technology can be reached. The idea that in-person hangouts with others are somewhat inherently better ignores the potential a digital-only friendship brings if the pool of potential friends is vastly expanded. One can get to know people from around the world, even if neither have the means to travel to each other's country. The technology necessary to enter and maintain friendships across the globe is generally ubiquitous, and thus the differences in availability of different kinds of people are growing between digital and analog friendships. Of course, we can seek friendships online and then make them analog, but for most people "available" online, this is not a viable option.

The fact that we can meet virtually anyone on earth, I believe, has not been appreciated in the context of the philosophy of friendship. The idea that we are somewhat limited to the people we are exposed to, or the rather limited technology available to make new friends (penpals, classifieds in newspapers), was taken for granted as a fact of friend-finding (see Sect. 7.3.1). Social technologies have not only expanded the horizon of public discourse (e.g., Twitter or Facebook) but also for private connections. If someone, then, chooses to become a digital hermit (or is forced to live as such), the options for them to make new connections are as varied and open as never before. The ability to get to know people from all walks of life, with greatly varying realities, assumptions about the world, and desires and wishes should encourage us to think that digital friendships will make those in them more worldly. If we take the idea that our local or present friends are a source of virtue or living a better life, we should welcome the ability to connect with a huge number of different people that will expand our view of the world.

Lastly, connecting to the last point, we may want to consider that technology has been a substantial benefit for many marginalized groups and their ability to connect with each other. Not only for a political organization or awareness-raising missions but also for the ability to connect with other people who share a common identity. The LGBTIQ+community, for example, has benefitted immensely from technology that makes participating in the subcultures possible: from sparsely populated to oppressive societies in which queerness in all forms is frowned upon or even prosecuted as a crime, technology enables people to connect. Obviously, this includes friendships. The ability of someone living in a highly homophobic society to talk about their identity with those living elsewhere in the world is an often overlooked, but highly valuable feature of modern technology. We should not assume that friendship between two gay men in Iraq is in any way less a friendship than a friendship between two men in Germany just because those in Iraq prefer to keep their friendship online.

The risks associated with being present with each other are substantial for many people, suggesting that only meeting online is the safest way of connecting with others.

6.5.2   Interim Conclusion—Being Friends with a Hermit

Hermitdom, as characterized here, can come in many forms. One can reject sociality in different forms, degrees, and with different levels of intensity. Ever since social media made it viable for people to stay in contact online to the degree that friends can, we have seen that friendships are indeed possible through technology alone. As others have similarly argued, we can be committed friends without having to have plans or wishes to ever meet the other friend in person. We can, according to Elder, even be excellent friends with each other as long as both sides are satisfied with the friendship.

We even have found arguments to claim that digital hermits and friendships built with hermits are likely to be better, as the ability to connect with anyone and for any purpose vastly expands the potential friends we have access to. Further, being able to talk or communicate constantly, and with fewer options to "spend time together", we can expect many of those friendships to be characterized by intensive conversations. Lastly, the opportunity for marginalized people to connect to each other, to find companionship, community, and a shared understanding of each other's life experiences points toward the strong good technologically mediated social connections, and among them friendships will certainly bring.

6.6   The Normativity of Immediateness, the Normativity of Opportunity

We can now conclude that friendships through technology are possible to the degree that non-technologically mediated friendships are, if not more. We have seen that technology is not only able to accommodate involuntary complications in one's life position, but that it provides even more opportunities to create friendships with practically anyone around the world.

We can claim that this marks a shift in the fundamental normativity of friendship-making. Never before was it possible for virtually everyone on earth to make friends with virtually anyone else. Of course, language barriers, access to (free) internet, and other practical restrictions apply, but even these last restrictions could vanish in the next years if we so choose (including the ability of machine translation services to translate our statements in real-time to the other person's mother tongue).

Thus far, the necessity of having to make friends with people around you has been accepted as a given, and thus people have developed expectations of being friends with those they meet in daily life. We can call this the "normativity of immediateness", in which both the norms of becoming friends, as well as the norms of remaining friends have been influenced by the given communicative horizon. This normative paradigm has been justifying the way we make friends, the expectation that we should be grateful for the friends in our immediate life surroundings and their potential quirks and idiosyncrasies. All these, viewed through the normativity of immediateness, help to find friends: learning to be tolerant and patient with others, valuing grown and older relationships over new ones and "origin" over "fit", sharing the same taught values—all those features support finding friends in one's immediate surroundings rather than attempting to befriend people who are closer to us in other features of our identity.

Social technology changed that. Not only can we become hermits and still participate in other people's everyday life, remain good friends with other people without having to leave our house or interacting with people we do not wish to interact with but also for the first time seek people we would otherwise never meet or hear of. The ability to seek people who share more of our identity, our opinions, and even simply our tastes, has shifted the normative landscape of finding friends: we moved from the normativity of immediateness to the normativity of opportunity.

The normativity of opportunity suggests that we should use the technologically provided chances to find people who are a good fit for us not because they are living in the same area around the same time, but because they actually fit better in our lives. Many of the criticisms of contemporary culture can be interpreted as a criticism of immediateness against opportunity: the supposed lack of tolerating other people's opinions, the creation of echo chambers or filter bubbles, the general overuse of technology, the loss of connection to one's surroundings—they all endorse the normativity of immediateness as something preferable over opportunities to seek people who are a better fit for us.

6.7 CONCLUSION: FRIENDS THROUGH TECHNOLOGY, FRIENDS WITH TECHNOLOGY

While we may take some of this criticism to be valid for the perception of nature, we take "opportunity" to be an underappreciated but ultimately decisive value emergent from the widespread use of technology. If we apply the expansion of the normativity of opportunity to technology itself, then, we may ask what the difference is between a digital hermit and a social machine. If the norm is not "to deal with what you are given socially" anymore but "to find your tribe", we should also consider whether technology can be the subject of friendship, rather than merely the facilitator of friendship.

Thus, we face the question of whether human–machine friendships are possible to the already minimized degree of being friends with an unembodied personal assistant or chatbot. The argument goes like this: if we can be friends with merely digitally present machines, we certainly can be friends with physically present machines, like robots. And if we can be friends with merely digitally present humans, why would we not be able to be friends with merely digitally present machines? This sets up the question for the next step of the investigation: We have laid the groundwork for asking whether we can reasonably expect to become friends with machines.

REFERENCES

Bailenson, J. N. (2021). Nonverbal overload: A theoretical argument for the causes of zoom fatigue. *Technology, Mind, and Behavior*, *2*(1). https://doi.org/10.1037/tmb0000030

Bowker, J., Matthew, H. B., Jonathan, B. S., Adesola, A. O., Rebecca, G. E., & Radhi, R. (2019). Severe social withdrawal: Cultural variation in past hikikomori experiences of university students in Nigeria, Singapore, and the United States. *The Journal of Genetic Psychology,180*(4–5), 217–230. https://doi.org/10.1080/00221325.2019.1633618

Chalmers, D. (2022). *Reality+: Virtual worlds and the problems of philosophy*. Penguin.

Cocking, D., Matthews, S. (2000). Unreal friends. *Ethics and Information Technology*, *2*, 223–231. https://doi.org/10.1023/A:1011414704851

Danaher, J. (2019). The philosophical case for human-machine friendship. *Journal of Posthuman Studies*, *3*.

Elder, A. (2014). Excellent online friendships: An Aristotelian defense of social media. *In Ethics and Information Technology, 16*(4), 287–297.

Elder, A. (2018). *Friendships, robots, and social media*. Routledge.

Fröding, B., Peterson, M. (2012). Why virtual friendship is no genuine friendship. *Ethics Inf Technol, 14*, 201–207. https://doi.org/10.1007/s10676-011-9284-4

Hamasaki, Y., Pionnié-Dax, N., Dorard, G., et al. (2021). Identifying social withdrawal (hikikomori) factors in adolescents: Understanding the hikikomori spectrum. *Child Psychiatry and Human Development, 52*, 808–817. https://doi.org/10.1007/s10578-020-01064-8

Kaliarnta, S. (2016). Using Aristotle's theory of friendship to classify online friendships: A critical counterview. *Ethics and Information Technology, 18*, 65–79. https://doi.org/10.1007/s10676-016-9384-2

Kempt, H. (2020). *Chatbots and the Domestication of AI*. Springer International.

McFall, M. T. (2012). Real character-friends: Aristotelian friendship, living together, and technology. *Ethics and Information Technology, 14*, 221–230.

Sharp, R. (2012). The obstacles against reaching the highest level of Aristotelian friendship online. *Ethics and Information Technology, 14*, 231–239.

Sherman, N. (1987). Aristotle on friendship and the shared life. *Philosophy & Phenomenological Research, 47*, 589–613.

CHAPTER 7

Synthetic Friends

Thus far, lots of effort has been spent merely laying the groundwork for what we can describe as the main chapter of this book. The main argument to motivate this chapter has been laid out over the previous ones: simply put, we found that historically and systematically, most philosophical approaches to friendship have been operating under a somewhat technology-free frame, and thus measuring psychological expectations and social norms against human nature and some cultural contexts alone. However, we suggested that we should account for different roles technology plays in our everyday life and that social media has made friendships possible that have not been possible before.

Further, we noted that some people prefer the lifestyle of hermits. While most philosophical theories reject hermits as potentially good friends, the differentiation between kinds of hermits has opened a way to bring together the novel technological aspect of staying connected without having to be physically present at all. We have also seen that we might have the intuition that seeing a robot in person for the first time after being friends with them can have a similar effect as meeting an internet-based friend for the first time. We concluded that the lifestyle of hermitdom is not precluding someone to become a friend, potentially even a very close one.

Thus, we have to ask the question we aim to answer in this chapter: what about machines entering this picture? We will discuss this question

H. Kempt, *Synthetic Friends*,
https://doi.org/10.1007/978-3-031-13631-3_7

in three different approaches: first, we ask if we can transfer what we said about internet-only "hermit"-friends to unembodied conversational machines. If this is possible, we have a strong case to consider machines friends in a very precise manner. In order to determine this, we turn toward some of the standard concerns about machine friends.

Second, we discuss what the benefits of creating those machines could be, and how these benefits can inform our opinion about human–machine friendships. This provides some productive insights into the desires friendships fulfill and how we should be confident that machines can fulfill these as well.

Third, we deal with what appears to be a substantial approach to human–machine friendship and contrast it with other, similar ideas, mainly Helen Ryland's approach to degrees-of-friendship approach.

7.1   What Is the Right Question to Ask

Hermits are not necessarily worse friends, once they decide and embrace the fact that one can be generally a "loner", an introvert with little regard for most people or society at large, and still be a committed, reliable, loyal, empowering, uplifting, loving friend. This is in part because technologically mediated friendship has allowed individual horizons to expand rapidly in a previously unprecedented way: what previously has been the phenomenon that we called "normativity of immediateness" has been replaced by the "normativity of opportunity".

These opportunities are to find new friends that align with one's preferences, identity, political or religious beliefs, hobbies, tastes, and desires. Instead of only having to deal with people in our immediate vicinity, we are now enabled to search for people of a similar kind virtually worldwide. This does not mean, and we will return to this rather simple insight in different places throughout this chapter, that the immediateness loses appeal, ought to be overcome, ought to be rejected, or otherwise frowned upon. For many people, friendships they make in their early years as children will last a lifetime. For others, people they meet in school, at university, at their first job, in their neighborhood, in their hometown or village, on their commute, via other friends are the ones that count; often enough people meet other people and befriend them because of a certain immediateness of their social connections.

What has changed and will only further change, is that our opportunities of making connections have grown. Our connections can become more intentional, more chosen, and deliberate. The question for any investigation into human–machine relationships, then, is whether or not machines will be able to perform the same way as humans do. Consider for this the following argument.

7.2 THE ARGUMENT
FOR HUMAN–MACHINE FRIENDSHIP

If a person now chooses to become friends with someone living as a digital hermit, thus refusing to engage in the usual levels of physical interaction (as we have seen, motivation for this behavior can vary widely), then they do not miss out on levels of friendship. They can have an equally fulfilling friendship with this person, potentially without ever meeting them. They can become closely integrated into their life, kept up to date on the most intimate concerns, issues, or wishes, and can provide advice and venting space (Premise 1).

If a person is not missing out on levels of friendship and is happy with their friends, then they live a good life with fulfilled friendships. There is not really more to demand from friendships than that, as the moral condition of virtue friendships is not exclusive to friendships, and other ways of seeing friendship can account for this perspective as well (Premise 2).

Thus, we should encourage them to form these friendships like any other friendships if it suits them. If we cannot recommend or encourage people to do what suits them, we cannot claim to be acting in the interest of people. It is thus required of us to recommend what we find to be suitable for them despite potentially opposing normative ideas (Premise 3).

If a person chooses someone like this (Premise 1), and if this person is not losing much in terms of friendship (Premise 2), then it is not unfeasible why a robot could not provide the same amount of friendship. However, if we can have similar friendships with machines as we can have with humans, then machines can enter these strong friendships (Conclusion 1).

If we can feasibly choose machines to perform functionally similar friendship duties to a digital hermit (Conclusion 1), and if we should encourage friendships like those with digital hermits if it suits a person (Premise 3), then we should encourage machine friendships to people if it suits them (Conclusion 2).

7.2.1   *Discussion*

I claim this argument to be valid and sound. Having laid the conceptual and normative groundwork, I do not see how one would insist that someone is having unfulfilling or otherwise defective friendships with machines without having to invalidate the lived realities of many teenagers, LGBTQIA+ folks in both liberal and oppressive countries, highly socially introverted people, and others. As a matter of fact, many people choose to have some of their friends as an online presence in their life. How the introduction of a machine into this lived reality would change anything remains unclear.

Of course, this is a rather functionalist view of friendship and human connection at large. But that has been established in the previous chapters on the argument that most human connections are now to be viewed from the perspective of technologically mediated opportunities. Having established this point with conclusions 1 and 2, we can turn toward what should be considered the main normative counter-argument against conclusion 1. Note at this point that we have so far said very little about the ingenuity of machine designs: all arguments so far and in the next chapter are purely based on how we interact with humans, and how those interactions are doing less work for the anti-machine-friendship arguments as proponents of those arguments suggest. The limits and potential of the technologies, and some ethical concerns about these, can be proposed at the end of these investigations to demonstrate the rather weak case anti-machine-arguments make.

7.3   COUNTERARGUMENTS

In the following, we will discuss some of the immediate arguments against human–machine friendships. Note that we have yet to say anything about the quality of the proposed machine friends, as we can assume that the following arguments will affect most versions of friends that are clearly on a conversational level, i.e., within the family resemblance of human–human friendships (rather than, say, merely human–animal friendship resemblances).

The main arguments discussed here are the ones about finding friends vs. creating friends, that human friendships are an aesthetic without which life is failed and that friends should function as an element of moral education.

7.3.1   Creation vs. Finding

A strong intuition about the difference between human friends and artificial ones is that the latter is created to fit a certain purpose, while the former is not. Stating this very fact is supposedly both a good argument against social technologies as friends, as well as an insight into the nature of human relationships. As Turkle puts it, "Technology is seductive when what it offers meets our human vulnerabilities. And as it turns out, we are very vulnerable indeed. We are lonely but fearful of intimacy. Digital connections and sociable robots may offer the illusion of companionship without the demand of friendship" (Turkle, 2011, 1). The argument comes in two parts, a positive and a negative one. The positive argument concerns the praise for the perceived arbitrariness of human connections, and our ability and necessity to make do with people in our life. The arbitrariness of human relationships is considered part of the relevant feature that allows us to form especially worthy ones (see, e.g., McKeever, 2022 for such a take on online dating). The negative argument consists in the rejection of artificial relationships as necessarily lacking the features described in the positive argument. Moreover, the negative argument is often accompanied by a similar, yet slightly different argument about the specifics of how we will create friendship machines and thus make it impossible for us to be friends with these machines. Essentially, the negative argument here states that (a) the fact that machines are created to be social is indicative of their lack of sociability and (b) the way we create their sociality prohibits the sociability required for meaningful human–machine friendships. We will first look at the positive argument in greater detail, to then analyze and assess the two parts of the negative argument.

7.3.2   The Positive Argument

This positive, often subtle or implicit argument has become one of the main ones because of two reasons: its intuitive appeal and the guilt we feel when exposed to such a judgment. The intuitive appeal consists in the fact that we perceive our friends as fully independent and distinct from us in the sense that it is a natural fit that we get along. Thus, our friendships are perceived as a certain arbitrary fit that makes them special. We, in this sense, do not pick our friends, and neither do they pick us. We meet by luck and grow together to become close friends by choice with a history that was only possible by being willing to grow in the first place.

Thus, the work we have to do in order to make friends is to be viewed under this context of arbitrariness. If we were clicking with others on a friendly level, the value of this connection lies in its indeterminacy and the work required to make a friendship out of it. Most of the philosophical literature is based on the assumption that creating, growing, and living friendships is character-forming and thus an integral part of becoming a well-rounded, virtuous person. Friendships, thus, are special because they happen only if we invest some moral capital in these relationships. We must be honest, interested, selfless, and open to someone to whom we are not related or have no other strings attached. This need, then, is presumed to be the worth of friendship only found in these arbitrary connections.

This interpretation of meeting people outside one's immediate connections, i.e., one's family, as the necessary element for having the opportunity to build a new connection largely covers our intuition of friendship.

With the lack of knowledge and opportunity to think about artificially created candidates to fill these roles, the negative argument rarely shows up in these deliberations. However, once we add technological mediation into consideration of human–human friendships even the positive argument will be affected by our ability to seek friendship more easily. Thus, we ought to consider the fact that social media has changed and will further change the conditions under which we seek and begin friendships with each other (cf. for an elaborate analysis Jeske, 2019).

7.3.3   The Judgment of Wanting Better Friends from the Normativity of Opportunity

Next to the intuitive fit of such characterization of how we make friends, there is a strong moral expectation that we should be happy with the arbitrariness of people's characters. Especially should we be forgiving with the limits and quirks of our friends, both because we are not better than them and because we should accept them as they are? Thus, forming the wish to want better friends should be motivated by severe misbehavior on the part of our friends. Any other reason to wish for better friends seems somewhat immoral.

We can trace the immorality of this judgment by two reasons: first, the arbitrariness of making friends implies that those friends do bring a certain conflict potential into our life. This may lead to disagreement, challenges, and other non-harmonious interactions every now and then (even if the friendship overall is rather close and harmonious). The opposite effect may be observed in the growth of filter bubbles and echo chambers online (Floridi & Chiriatti, 2020), in which people group themselves according to the confirmation of their political views online. In the mid- to long term, these echo chambers will reinforce the own political position as correct, thus causing people to lose the nuance of political discourse. Friendship, in this sense, then is often precisely not an echo chamber, but can be a test for our ambiguity tolerance and thus encouraged. We can imagine two friends who intensely fight over the role of unions in the economy, with one being more pro-union and the other pro-business. With intense, emotional debates, while in the end still being able to go to a bar afterward and spend a good time together.

Wanting to decrease the amount of resistance we may face for some of our political, cultural, or other views may be helpful to keep us open-minded to other potential conflicts as well. Thus, the arbitrariness of friend-finding somewhat keeps the chances alive that we may disagree with our friends. And disagreeing without friends guarantees that we can guarantee with strangers about political issues as well, benefitting political discourse all around.

Wanting to have "better" friends, in this reading, usually merely means that we want friends who are less challenging than what we want ourselves, and thus the desire is an egocentric one. Of course, we can imagine unfortunate situations in which a genuinely good person only meets genuinely bad people, thus only having bad friends. This person's desire for "better friends" should not be condemned. However, as far as the argument here of the desire for wanting better friends goes, this is not about replacing one bad friend with a better person, but rather wanting the same people to just be better friends. If it really was just about having friends who are better friends by virtue of being better people, we could seek different people rather than remain with the ones we have.

It seems that the second issue, the guilt of wanting better friends, is supposed to be more telling about our morally depraved or otherwise rejectable urge to create friends that are more "suited" or harmonious to our needs. This judgments of "guilt in wanting better friends", I think, can be traced to two different motivations: either, the "natural-ness" of finding friends will create relationships that are naturally prone to disagreement, disappointment, challenges, and other non-positive, non-harmonious interactions. Or, the result of those "natural" processes of finding friends will lead to friendships that in their content and quality will add to a wider worldview.

We can have a roundabout understanding of what these types mean: the first kind of relationship brings together people who will clash, have issues, but are still full of love or affection. The second can be, on a personal level, harmonious, but bring political, cultural, generational, or otherwise stratified, collective differences into a relationship. However, the reason why the "naturalness" of these types of relationships ought to be preferable over "artificial" is the requirement they put on us: given that humans will have conflicts with each other, we must learn to tolerate those conflicts as part of our social connections. However, not only should we tolerate those conflicts, we should appreciate that the other person's perspective is one to be taken into account when attempting to resolve these issues. We may find them egocentric or otherwise inappropriate at times, but the fact that we have connections with people who want things independent of ourselves requires us to seek compromises with others.

7.3.4   The Problems with the Positive Argument

The positive argument for the "natural" way of finding friends suggested that this is a natural way to make friends and that those friends, due to their arbitrary background, might help us become and remain more tolerant people all around. And while these claims may hold anecdotal evidence, whether or not personal experience with culturally or politically varied opinions will lead to a bigger tolerance of those views remains to be seen (and even properly evaluated, as we should not want people to become more tolerant toward intolerance, merely because they were exposed to it more). However, next to the dubious empirical claims being made here, we have to discuss two arguments that speak against the positive argument. The first one regards the quality of those arbitrariness contexts, and the second is whether we have allowed arbitrariness to rule our friendship-seeking before.

7.3.5   Growing Up, Growing Out

First, not every experience of the "friends by randomness" is a positive one. Friends made during childhood in the immediate neighborhood or elementary school may last a lifetime, but also may not last at all. Depending on personal development and life choices and preferences, friends who spent the first few years inseparable may end up never talking again after going to different places to study. It is not unlikely that childhood friends could end up on the opposing sides on some cultural or political issue. Thus, we should be careful with romanticizing certain "analog" friendship building structures as "the proper way" to build friendships, because this may not benefit everyone involved. Especially in structurally conservative or oppressive contexts can personal developments mark the end of a friendship.

LGBTIQA+ folks, for example, may perceive the hometown where they grew up very differently from their former best childhood straight friend, who still fits in those small town norms and expectations, while the former is not. Perceptions of childhood, one of the elements of shared memory and identity, can vary vastly depend on the development of features of personal identity. This shared identity of being from a certain locality, having spent afternoons or summers in certain places of that locality, being familiar with the history, mythology, and gossip of that locality is not necessarily the identity-motivating marker that people think it is. Growing out of these contexts often means losing the shared identity attached with those childhood places. Of course, this does not mean that most people will reinterpret their shared memories and thus lose their shared identities with childhood friends. It might even be the case that many people remains friends with their childhood companions precisely because these shared memories retain strong sentimental value. It would be mistake and ignorant of many people's lived realities to elevate these arbitrary childhood friendships with whom we share lots of memories to prove anything about friendships in general.

In this light, it seems unclear why the arbitrariness of making friends is a positive argument in itself. Often enough, the friends we make randomly are not the ones that stick around, and that for good reason. We grow as persons and thus develop in ways that may not work for childhood friends unless they grow with us. Those old friendships may retain sentimental value, but why these should tell us anything about friendships remains unclear. Further, the next argument will elaborate on the fact that we do, indeed, choose and sometimes even abandon friends based on our preferences.

7.3.6   Fact-Checking Friend-Finding

Second, however, is a much more consequential reason that should be spelled out in more detail. That is the simple claim that most often we do not, in fact, find friends in an arbitrary context. For this argument to fit better, we should elaborate the claim a bit more on what is meant by "arbitrariness": what Turkle and others presuppose about the praise of the natural encounters and the real people we make friends with is that they are complex, with their own agendas and interests, and thus not a comfort to our life, but something we have to adjust to (of course, this is a more powerful argument with the contrasting claim that human–machine friendships would only be a comforting, narcissistic tool to feed an ego).

However, once we look closer at the way how many people meet people, the scope of the presumed arbitrariness is shrinking fast. We pick friends with some kind of criteria by way of meeting them: going to a sports club, a political party, a special university event, or a church implies at least some kind of shared interest we exploit; we also go by sympathy, a more subconscious yet still a somewhat controllable feature of friends.

Now, we might want to say that we still just meet people there, being exposed to the arbitrariness of life and people's pre-formed personalities who are not there to simply please us. We still have to "make" friends and withstand their personality quirks. In short, we need to accept the arbitrariness of all their character traits, even if we are attracted to some of those traits.

But it already is less arbitrary than meeting people on the streets or blind. Usually, there is a connection moment that makes them sympathetic or funny, which makes us want to be perceived as sympathetic or funny as well. And there is a chance that the other person also tries to be perceived as such. Thus, the effort someone makes to appear sympathetic by looking for bonding moments should be discounted from the arbitrariness factor. Wanting to be friends with the cool kids at school, the witty office coworker, or the neighbor with the amazing travel stories, are all impulses based on the sympathy someone has for someone else. We can even go further and assume that most people would reject others even if they appeared sympathetic if their political beliefs did not align at least to some degree. Without evaluating this point, we should acknowledge that friendships are often conditioned on our personal preferences about other people's worldview, attitudes, or even simple traits. It thus is factually false that we make friends based on features that are indifferent to the later friendship.

Lastly, even before social media was created there have been events and even clubs specifically for connecting people on a friendly level. That means there were and are events to which one can go to make a friend. Of course, it is not a pick-and-choose kind of activity, but most social networking events that do not have a specific career dimension to them are, in a way, friendship-production events. This does not invalidate the fact that friendships do happen to form arbitrarily between people who just get along based on a few features. However, once we acknowledge the intentionality of exposing oneself to certain friendship-finding-contexts, or the explicitness of finding people specifically sympathetic and wanting to be friends with them, this thesis and thus the positive argument of friendship-forming decreases in its appeal and plausibility. The fact that people do, in fact, expose themselves to those environments could rather be interpreted as evidence for the intentionality of friendships.

7.3.7   Should We Want Better Friends?

Seeing now that the positive argument relies heavily on assumptions of contingency and romantization of conflict, we are better equipped in responding to the guilt of wanting better friends. This guilt does not stem from the arbitrary contexts in which we supposedly make friends, as these contexts are indeed less arbitrary as often portrayed: we put ourselves into rather intentional positions to meet people we have reason to assume have similar interests to us. We meet friends through other friends, sometimes even being referred to others. We meet new friends in sports clubs, on websites and social media apps, and at work. All these suggest that we meet people, not at random. Sometimes we even pick people specifically and try befriending them. Friendship is only arbitrary in a very limited sense, if not even less so.

What does the feeling of guilt then tell us, if, conceivably, most of our friendships are indeed partially intentionally entered in the first place? Why do we not want a better version of these friendships, or rather have moral scruples to wish for them?

Certainly, there is self-awareness at play in which we do not want to be thought of by our friends as replaceable or improvable. While we might know the feeling that we wish a friend was more interested in something we are interested in, wishing for this person to think of us the same way can be hurtful. Friends as a "given" social connection that we usually do not regret is based on the mutuality of the reliance on the connection. However, some friendships are more intimate and precious than others, and it is not inconceivable that some friendships are indeed becoming more burdensome.

For those friendships, in which we might not feel seen, or in which our friends are disappointing, or in which, maybe, the unspoken ease of connection is disappearing, it is hard to see why we should feel guilty in turning our attention away. We do owe friends our loyalty, but it would be uncalled for to require the same unwavering loyalty from every friend in every instant. After all, we might just want different friends at some point in our lives, and we should accommodate this wish as a representation of personal growth that not always is linked to the growth of a friend.

Thus, we can distinguish between purpose-driven friendships, for which we might not feel guilty (or should not feel guilty), and those friendships, which we call irreplaceable. For the former, we can conceive that these friendships could be different, more intimate, and more aligned with our interests and character. For the latter, this is not the case. Social connections have a certain path dependency, and one could claim that the described guilt is addressing that.

This could remind us of Aristotle's approach of distinguishing friendships (Chapter 5), as we might want to call the latter virtue friendships and the former utility and pleasure friendships. However, this is inaccurate. We could fear replaceability for someone we merely seek for mutually pleasurable experiences, as those connections may become very important to us, too. As Aristotle is judging from the moral character of those friendships and not the amount of pleasure realized, we can still fear for our lesser friendships the same way we should be concerned about losing out morally worthwhile ones.

7.3.8   The Negative Argument

The argument for the natural arbitrariness of meeting people stands in contrast with the condemnation of humans seeking mere validation, pleasure, or confirmation in human–machine friendships. Thus, so goes the rather straightforward argument, we cannot be friends with robots in this philosophically more demanding way: not only do we miss out on the positive effects of friendship-finding but robots are the worse option as they do not allow for these kinds of connections. As mentioned before, the positive and negative arguments combined usually represent the full argument authors make about the issues of intentionality of design in human–machine friendships. However, as we have seen with the positive argument, there are reasons to analyze the negative argument separately.

The negative argument claims that we cannot create a robot that brings the independency and ambiguity with it needed to count as a proper friend. Even worse, as we would not create or interact with a machine that is frustrating us in its behavior, we should not expect machines to behave in these kinds of ways. However, without these independent, self-assured perspectives that a robot simply cannot have, we should not expect robots to become able to enter human–robot relationships.

In order to tackle this argument, we first are going to elaborate more precisely on what this charge means, to then argue first that, coming from the last point of the positive argument made about seeking friendships. Here, we can see that there is a slippery slope argument at play where we disagree about where the slope goes. While the negative argument discussed here suggests that allowing human–machine friendships to count as full and proper friendships will lead to a further disconnect between actual human beings, we interpret the slope to go down the other way: if we reject human–machine friendships based on these naturalness arguments of human–human friendships, we will be forced to reject some human–human friendships based merely on their lack of naturalness, or because of the intentionality of connection.

Further, we explore if there might be salient, cultural reasons why people reject human–machine friendships. If so, we have reason to believe that we should discount the strong skepticism about the deterioration of human–machine friendships as mainly motivated by generational concerns. We will incorporate Gunkel's and other's concerns of "othering" as a cultural exclusionary process and show that this is sufficiently similar to exclusion practices.

Lastly, we propose a productive understanding of this concern that can inform future designs of machines capable of entering friendships.

7.3.9   *Rejecting Robots*

Robots are manufactured and designed to serve a purpose. As we have seen from Chapters 2 and 3, the construction of such machines is well-situated within business models with an incentive to keep you happy with the product. Thus, in opposition to what human friendships usually do, human–machine friendships will necessarily be limited. This, as well as the positive argument, appeals to a certain intuition about friendship in which friendships in which friends always behave exactly as we want them to be defective. It appears that the unease we may feel when thinking about human–machine friendships is motivated by this fact: if we are realistic about the prospects of establishing friendships with machines, we should see that an entity without personal interests will not push toward some of the more important markers of friendship: challenging us to care for someone for their sake; respecting boundaries, preferences, emotions, someone's past, their vulnerabilities and traumata, etc. These can, reasonably speaking, not be recreated and at best be imitated. The large vault of biographical, personal, psychological stories we can tell about our friends, especially the close ones, will not be replicated within the near or even plausibly the mid- to distant future. We would have to achieve the so-called AGI, artificial general intelligence, to create a consciousness that could provide the space for developing those traits, and even then it is not guaranteed that we would actually be able to create something akin to a friend.

The underlying but seemingly justified premise is that humanity has put itself in a privileged position to enter friendships, rendering attempts in adding robots futile based on the very idea that these intense interpersonal moments do, indeed, require persons.

This demonstrates the rather simple statement that machines will not be able to take part in friendships due to the (a) lack of their own personal psychological stories we could relate to, and (b) their lack of genuine interest in our psychological stories and biographies. Consequently, we should reject technologies that pretend to care in this regard, as neither the recipient of our sorrows genuinely care, nor has this recipient anything to offer on their own that we could relate to in any kind of sense.

Further, the very fact that we create machines that use the language we use to express our care for others, or the language we express our own need for being listened to and cared for, can be offensive. People may reject technology that entertains that very fact without being able to follow up on the implied care associated with "lending an ear". This offense can be translated into an argument that social machines of this kind should not be allowed to participate in what they perceive to be exclusive verbal jurisdiction of human agents: machines should not use human ways of speaking to "calm" or "relate to" or "sooth" a human agent in pretending that they in fact cared about the problem in question. This might be the most consequential result of assessing machines using human speech: if we allow machines to "pretend" to be human, in even the more obvious imitations, we may enter a slippery slope in which we end up at some catastrophic outcome.

7.3.10   Slippery Slopes

The presumed slippery slope looks like this, roughly. We have several stages of more or less acceptable steps of technological development. Every little step of development brings us closer to a certain unwanted state of affairs, but no step on its own is particularly problematic. The catastrophe for people arguing along this slope is a fully integrated human–machine relationship that might replace some of our actual human–human relationships. What makes this state of affairs catastrophic is often hard to explicate.

Once we notice that we are on such a downward slope toward proper human–machine friendship, people begin searching for identifiers of where we "went wrong" or "will go wrong" in our cultural or psychological development to warn us from further sloping downward. From these points onward we can identify some stages in which this slope becomes apparent and, once we agree that we do not want this particular outcome, we are called to intervene at earlier stages of this development (even when the particular development is morally innocuous). One of those early stages that could be considered is the fact that we train machines to talk "like us". Anthropomorphism in itself is rarely considered morally problematic and often merely a starting point for research on recreating human intelligence or creating new social intelligence. The first steps in anthropomorphism are barely visible as such and are often necessary to make a system operable in the first place. Anthropomorphism comes in small degrees: from merely using human words and potentially human phrases to fully simulating a personal mental theater, the question of whether a design is inspired by human models or whether certain features are unavoidable no matter the fact that the question of whether the inspiration is taken from humans can rarely ever be answered. The problems of anthropomorphism as a fundamental design choice are unavoidable, thus the slope is slippery.

Another moment on the downward slope could be the construction of social machines even if not intended or designed for more private or even intimate human interactions. Social machines, as we have seen earlier, are often designed for specific social contexts. Even a personal assistant on our phone that could be interacted with all day long does not suggest itself as being a "friend", as it does not initiate conversation or ask questions to entertain us. However, even there people have begun using the social interactiveness of these assistants to enter some kind of bonds or attachments. A similar example would be the use of audio signals of certain machines to imitate human emotions (a sad "beep" will trigger an empathetic response from people).

The very fact that we can have any kind of stimulating or socially satisfying interaction with machines could be interpreted as yet another step toward the replacement of human intimacy with some kind of factory-made, illusionary intimacy that does not challenge us to become better people by caring for others. If we take the slippery slope argument seriously here, establishing these innocuous instances of human–machine interactions, the improvement of those interactions will create eventually human–machine relationships that are a danger to our human intimacy.

Against this interpretation, however convincing it may seem at this point, we should provide an alternative interpretation of what is going on in these slopes. The issue with slopes, usually, is that we cannot reasonably find a stopping point for the development that is deemed unproblematic at the start. However, if we consider the end (here: human–machine friendships) to be rather unproblematic it becomes clear what kind of relationship of humans with technology has been deemed unacceptable. From this perspective then, the catastrophe described in this slippery slope does not consist in the establishment of unacceptable human–machine friendships, but rather in the ever less tolerant attitude toward technology-mediated friendship. As we have seen with digital hermits, we worked toward decreasing the concern of this argument: at this point of technological progress, many people have friendships that mainly or even exclusively are being conducted online. Yet, we would not judge them to be worse friends on mere moral grounds. If we reject human–machine friendships on the basis of something missing that cannot be replicated, we might see other human friendships as potentially equally defective.

The slippery slope in this regard begins with arguing against human–machine friendships: if we reject those kinds of connections with machines, we have to reject some of the more technologically mediated friendships as incomplete or defective. If we have to reject those severely technologically mediated friendships as incomplete or defective, there is little leeway in declaring most friendships that function on the basis of technological mediation as defective (even pen pals). The catastrophe would be that we would be required (as dictated by intellectual thoroughness) to take the position that we simply cannot know at this stage whether the careless video calls with someone's grandma who lives across the globe will not lead to normative troublesome issues. This is because we rely on technology that maybe, down the slope, will replace human intimacy with an artificial one.

This demonstrates the absurdity of reading of the slippery slope argument against human–machine friendships. Additionally, we have not even mentioned the fact that this argument might even be overreaching in its scope, since it also presumes that human intimacy will be weakened through the rise of human–machine friendships. This is a particularly culturally pessimistic assumption about the willingness of people to engage in social connections and their preferences.

We endorse the view that the slippery slope, if anything, is motivated by arguing toward a more free society, and that if we rejected human–machine relationships, we would face serious concerns for the evaluation of proper human–human friendships where both parties prefer remaining online. Digital hermits, in this sense, ought to be a moral reason in favor of human–machine friendships. Otherwise, human–machine friendships will be a moral reason against digital hermits.

7.3.11   *Friends Are a Moral Entryway*

Related to the authenticity- and arbitrariness-requirements of friendship we can carve out one specific dimension of human–human friendships that also has been mentioned in the previous chapter. Friendships require moral behavior and thus build an ethical character, and human–machine friendships cannot achieve that.

This argument can be related to the hierarchy-conception of friendship we had discussed in Chapter 5. The assumption about this typology of at least three different ideal types of friendship was that to participate in the highest form, the virtue friendships, we have to leave the logic of exchange of goods or pleasures. We have to become, in some minor form at least, virtuous by being genuinely interested in the success and well-being of the other person. In order to achieve that, we have to recognize that others may have goals different from our own and that helping others achieve their goals potentially to the detriment of achieving our own is a worthwhile thing to do. Behavior of this kind in friendships is supposed to be more praiseworthy than with family members because of the lack of a biological connection. There are no family- or other instincts that would compel us to act altruistically. Friendships, as a voluntary connection between two unrelated people, are thus the best opportunity to demonstrate one's virtue as one can prove themselves capable of taking someone else's perspective and act according to their interests.

At the same time, making friends virtuously and thus participating in this ideal type (to whatever small degree) teaches us that other people are also capable of acting virtuously. This reinforcing effect of being virtuous and being treated virtuously can establish a moral character. In fact, Aristotle himself considered friends the key for developing a virtuous character.

The argument here, then, for rejecting human–machine friendships rests to a degree on the assumption that the philosophically relevant friendships are only those of the highest regard. Nyholm (2020) even admits as much: "To be sure, robots can be useful for people. And they can also give us pleasure. So at least on the human side of the equation, we could say that robots can provide us with some of the goods we associate with having utility or pleasure friends".

Virtually every philosopher involved in this debate takes this argument to be valid, and as far as I can see only Danaher argues against it while everybody else works around the consequences of such argument (with Ryland avoiding the question of classic virtue friendships altogether). Machines, neither having personal interests nor a moral character to build on, simply cannot be virtue friends, and thus we cannot become more virtuous through them.

We can present three arguments here rejecting the consequences of this argument. The first one was been presented by John Danaher himself. As he argues, we can break down the concept of virtue friendships into several conditions and test whether these conditions can be fulfilled (see Chapter 5). Merely the metaphysical condition of personhood/moral subjectivity remained a challenge, while all other conditions could be resolved through technological progress. Once we add Danaher's theory of ethical behaviorism an argument forms in which the programmable ethical behavior of a machine can be on par with the ethical, virtuous actions of a human. Thus, machines could eventually fulfill all conditions required to achieve this goal and thus contribute to humans becoming more virtuous friends.

Sven Nyholm's argument against such progress lies mainly in adding a fifth criterion to the mix which, as he claims, cannot be rejected within Danaher's theory: the ethical condition of whether such machines should exist in the first place. As Nyholm doubts that the technological solutions reach the conditions genuinely, ethical behaviorism essentially also requires to justify the production of these machines as well. We should, according to Nyholm, not accept these machines, as some of Danaher's own conditions, i.e., equality, must mean "the equality in rights and moral status" (Nyholm, 2020). Thus, this might constitute a violation of the fifth condition, and thus the goal of creating virtuous machines becomes an inconsistent goal (as the realization of two conditions conflict with each other).

The way we have set up our investigation should suggest that we can indeed answer the ethical conditions independent of ethical behaviorism and thus reject Nyholm's objection. This ties into the second argument we can discuss here: it concerns the assumption that virtue friendships must be equally both sided. However, we, as human agents, can become better friends by experiencing virtue-friendship-like situations with machines. This argument may not resolve the issues if we actually enter human–machine relationships, as it does not give us a reason to believe that virtue friendships with machines are possible. However, this was not the thrust of the argument: the point of virtue friendships was that we would be enabled to become more virtuous, and a certain virtuous vortex between two virtue friends would make both better.

Now, entering a virtue friendship with a machine that gives a rather convincing performance of a virtuous person may be violating some of the conditions of virtue friendships, such as the metaphysical condition of personhood. However, as far as the human side of such friendship is concerned, they can still grow to become a more virtuous person. This could be seen as similar to owning a pet to instill a sense of responsibility in kids. We do not have to assume that a hamster is actively depending on us and would blame us if we forgot to feed them, yet the felt pressure to be responsible and reliable in owning a pet instills in us forges our moral character.

A machine with whom we hold intense conversations about the right thing to do could still compel us in aiming to do the right thing, to improve ourselves, and even to forge a better moral character. In this sense, Nyholm's concern about the fifth, ethical condition also appears less convincing: why should we be concerned with machines that enable us to become more virtuous?

Friends may be a moral entryway because they compel us to genuinely care for people who are from the same family like us, thus building a better community. However, so far we have not seen a good reason to reject machines as potentially helpful elements in this attempt.

7.3.12   Friends Are an Aesthetic

Another argument against our core argument of human–machine friendships concerns the fact that next to the authenticity of friendships, the conditions of being virtuous friendships, and the similar life experience conditions, friendships are aesthetic. As we have discussed previously, this idea has been proposed by Nehamas, who described friendships not as a specifically ethical endeavor, but as one of the aesthetic quality of life. Those who want to live a life of the highest quality, then, ought to seek friendships that contribute to this quality. This renders friendships an aesthetic element in a successful life. This approach is more comprehensive to capture the value friendships have in our lives, as we usually do not see friends as a mere source of ethical motivation or representing different parts of the three idealized types of friendships, but simply as sources of quality of life.

In this sense, we could judge machines simply not being part of this aesthetic assessment and thus not contributing to the quality of life as imagined by Nehamas. Machines in this view are inherently less "aesthetic" than humans in their friendships. They do not love and admire and thus are open for all kinds of experiences. This ties up most of the points made so far in this chapter: machines may merely simulate excitement about our lives while genuine excitement is needed for such aesthetic quality. Machines do not relate to us as we relate to them, while human friends relate to us from a shared base of memories, experiences, and inner processes such as emotions, even the "merely digital" ones. Machines, at best, merely "make us better people" which does count toward their aesthetic quality; but not to the extent to which a human person contributes to our life. They merely, so goes the argument, remain an addition or a feature to our life rather than part of our social infrastructure.

I would argue that we can both consider this true and yet not affect the core of the argument. There is at least some aesthetic quality to be detected in human–machine friendship: they can contribute to lasting experiences, encourage us to try new things, to develop as a person. Arguably, in a mediated sense, human–machine friendships can help us lead better human–human friendships from this perspective as well. If, for example, someone would like to be a different kind of person (try a different hobby, dress differently, etc.) but is afraid of exposing themselves to anyone who may judge them, a machine can help provide non-judgmental feedback, and with that build self-confidence to expose ourselves to friends. Ideally, of course, we would expect our friends to be open and non-judgmental already; however, the step between the utmost private and the social could be mediated by a machine friend.

7.4   Some Constructive Proposals

For most of the investigation so far, we have dealt with defensive arguments: is an investigation into human–machine relationships even justified? What are the conditions of creating social machines, and are those conditions not prohibiting any reasonable attempt in creating machines to be friends with? How come the history of philosophy appears to unprepared for the rise of digital-only friendships, and does this affect our ability to be good friends in the future? And why should we take an argument in favor of human–machine friendships seriously if so many strong reasons speak against it?

This is motivated by two reasons. First, the risks currently outweigh the opportunities. We are simply not in a position at the moment where a productive proposal about the contents of human–machine friendships is a realistic endeavor. Rather, philosophers have been busy outlining the very potential of human–machine friendships and aimed at finding the limits and boundaries of such an undertaking. As we have been developing the conditions for such undertakings in Chapters 1–3, we found that there are concerns about the social ontology, the production conditions, and potentially even social acceptance and difference that will influence the success of those social machines or lack thereof. That philosophical inquiry currently is concerned with arguing about these boundaries and conditions suggests that most have a realistic understanding of the current states of technological development. Second, more importantly: once we start making an explicit positive proposal about the behavior of machines for friendships, we are reinforcing some of the arguments about arbitrariness. Creating friendship machines seem problematic, creating machines that happen to be able to connect with us on a friendship level, however, do not. The idea that we should give positive impulses on what friendships with machines should look like, rather than merely giving guardrails on how to avoid exploitation and other harms, would show exactly the creation of a mere friendship machine. We should, instead, encourage human–machine friendships to be as varied and multi-layered as human–human friendships are, eliminating the need to give specifics about machine design to fulfill these friendship conditions.

Do these two reasons suggest that there are no comments to be made on the content of such friendship machines and the potential human–machine relationships? At least it suggests that those comments should be minimalistic, and we should be careful with content proposals. It would be a mistake to declare any kind of specific human–machine friendship better or worse than others given the fact that the conditions and limits presented before are observed. Not every human–human friendship is the same, and they seem to differ along many factors that are up to the people in those friendships alone.

All that has been said in a similar fashion by Helen Ryland that we have discussed above (Chapter 5). However, going beyond her proposal, we can think of some positive features that would not only help machines become friends, but that replace intuitions about other positive features, such as anthropomorphization.

7.5   Conceptual Interlude: Are We Still Talking About Friendships?

One worry from this development could be that we do not talk about the same concept anymore. Ryland, Danaher, and others stated that they wanted to add machines to the mix of potential friends, rather than built a different concept altogether. However, as we have understood friendship to be merely a concept with family resemblances, and if we agree with Ryland that necessary conditions for identifying kinds of friendships are rather problematic, while also agreeing that her solution does not get us there, we can make a new proposal: Human–machine friendships can and should work fundamentally different from human–human friendships, though with similar features to them.

It is thus not about the recreation of friendship that has been common so far among humans, but the creation of genuinely new relationships with machines that can be called friendships, since they family-resemble classic human–human friendships. This can stretch and change the term; however, as we have seen with other relationships and other types of friendship (e.g., to pets), the evolution of certain relationship features that are complementary rather than replacing (see also Darling, 2017, 2021; for an account against such comparisons from machines and pets see Johnson & Verdicchio, 2018). Thus, we should feel confident that the idea of human–machine friendships as a new kind of friendship is not in reality a different kind of relationship for which "friendship" would be an inadequate term.

7.6   Synthetic Friends, the Very Idea

The very idea of synthetic friends, then, is the following: in attempting to reconstruct friendships with machines to embed them with the concept of friendships with humans, we make a mistake. Ryland's attempt in reducing it down to "mutual goodwill" still requires a certain subtle mental imaging that we better avoid. We thus should offer some conditions about the technology, and ourselves, that are necessary for something to be reasonably called a friendship. Recall that we just stated that the term friendship, due to its generally ambiguous nature, should usually include an addition to include "friendship between whom"? This marks the distinction between kinds and degrees of friendship.

For a friendship between a human and a machine, then, we should require different conditions than between two humans, as they might as well merely have to exhibit goodwill and another feature to be considered friends. It also allows for other definitions of human–human friendships to be considered or held up without interfering with the idea of synthetic friends presented here.

Before we should elaborate on those conditions, we should explain the choice of terminology. The term "artificial" has wide spread use, from its main application in AI to other established techno-philosophical terminology like "artificial moral advisors" (Giubilini & Savulescu, 2018). Artificiality however implies a certain recreation of something natural:

artificial intelligence is to be seen as a counterpoint to natural intelligence without questioning the "intelligence"-part. Thus, artificial friends, meaning machine friends, would suggest that we merely recreate friends in an artificial way rather than create something genuinely new. Further, the normative associations one can have with artificiality should be avoided as well. Artificial friends may sound like fake friends, rather than a new kind of friend altogether. And we can be in human–human friendships that turned out to be highly artificial, or fake.

Synthetic friends on the other hand suggest that there is a manufactured dimension to these friendships, but not that they have to be in any kind fake or merely a recreation of something that can be found in nature. It also suggests a difference to other family resemblances of friendships, such as human–animal friendships, or friendships among football fanbases. This way, the term may contribute to the debate about friendships as it can be integrated into the wider spectrum of friendships.

7.6.1   Conditions

In the following, we will present and defend the conditions necessary for human–machine friendships to emerge and last. We should consider this list exhaustive, as every condition is necessary but no other is.

7.6.2   Relatability—Social Technology

As we have seen in Chapter 4, we should only consider technology that fulfills the conditions of being relatable, i.e., relational technology. We cannot be friends with something that does not exhibit some of these relational technologies' features, which are most importantly autonomy, attachability, and interactivity. In the following, we will shortly recap these three conditions to then argue why relatability is a necessary condition for human–machine friendship. We name these conditions specifically rather than merely pointing toward "social technology" for two reasons. First, some social technologies work are mediators of human interactions rather than relatable technologies. This means that a social media platform, for example, is a social technology but not sociable. The term "social technology" as a subject for sociological, psychological, philosophical, and other research purposes, thus, might not be specific enough for our purposes here.

And second, there is also a positive difference between the set of technologies that fulfill these three necessary conditions and the set of technologies that count as "social technologies". We can, e.g., relate to technologies that are not intended to be related to. Technologies that end up playing a social role and social technologies, thus, are two different things, and to avoid excluding some technologies that could end up playing a social role for someone, we should go by some necessary conditions only.

Autonomy is often considered a requirement to perceive some technology as more than merely an unrelatable thing. An autonomous machine is one that behaves with a certain independence from human input, and thus with somewhat unforeseeable moves, even within certain margins or with hard limits. This is a necessary condition for any technology to appear to us as if it is acting purely on its own.

Thus, with autonomy, we usually mean "acting on its own behalf", even if the behalf is pre-determined by some engineers or other external forces. The ability of a machine to pursue its own pre-established goals is a key element in perceiving technology in a certain relatable way. Without the other necessary conditions, autonomy on its own is insufficient. Take, for example, an autonomous surveillance system that is programmed to turn on and record when being triggered. This is, technically, an autonomous system but in no relevant way relatable to us.

Interactiveness is an additional, yet necessary condition for relatability. With interactivity, we usually mean the ability of machines to react to our input, while we can adjust and align our actions with the machine's behavior. Thus, it is a back-and-forth between human and machine behaviors. The reciprocal alignment (Branigan et al., 2010) of machines and humans has been researched strongly and works as a general condition for relatability. Interaction, in this regard, can come in a lot of different depths, as philosophers have begun distinguishing between different kinds of interaction, from "teams", "collaborations", "cooperation", and "assistance". These interactive machines, however, are not necessarily relatable. An interactive toy, for example, a video game, is not relatable in this sense as we do not build a relationship with the game itself, though arguably that is the object with which we interact. Thus, interactivity alone cannot be the source of relatability. Even autonomy and interactivity combined will not achieve the necessary conditions: we can imagine a video game that works fully autonomously and with which we can interact, but that does not provide anything for us to relate to still, e.g., a random number generator for certain single gameplays.

Attachability might be the most ambiguous of these three conditions and as such should be explained in more detail to ensure that this necessary condition is not overreaching. The concern for the ambiguity mainly emerges when trying to unpack what it means for a technology to be "attachable".

On the one hand, it implies that the technology can attach and connect to our human activities, on the other hand, the technology should provide some surface for us to project some wishes onto. Thus, attachability is both a feature of the machine attaching to us, as well as a feature of the machine for us to attach to. The latter is presumably what Gunkel means with the "face" of a machine. Not meaning a literal face with eyes and mouth, but rather the qualities of a machine to be perceived as an individuated thing that interacts with us, rather than anyone. We can "see" the machine as something that can become part of a social context, as a character or something with a "face".

What does it mean for a machine to be attachable in the first sense in order to be relatable? We can describe this as a certain quality of interactivity and autonomy, in which the behavior of a machine attaches to the norms of a certain social activity, rendering it un-attachable. There must be, in this sense, a certain behavioral base that resonates with us and our norms of behavior. A machine that merely does its own "thing" in its interaction with us, which does not mesh with our goals or seek human interaction though it can interact with humans, is not in this sense attachable.

On the other hand of the attachability consideration stands for the idea of a "face" or some other characteristic that allows us to identify the machines as an individual to which we may built a social relationship. This often translates to specific anthropomorphic conditions, like an actual humanoid form (bipedal, arms, a literal face with eyes), though this is not required for a machine to have a face in this rather metaphorical sense to being attachable to.

We can also reintroduce the idea of pragmacentrism, in which the participation in discourse was considered the relevant feature for moral and social consideration. We will explore the idea of embedded social relevance as a marker for social consideration in the next chapter, as friendships often come not only in isolated conditions but in integrated contexts. The implications of pragmacentrism, then, could be useful in assessing a machine's ability to serve as a social relational object, though we may want to grant that exhibiting discourse participation behavior (see also ethical behaviorism) is too much for the necessary requirements.

The other conditions of these lists are phrased in the negative. This is due to the explicit rejection of any positive necessary condition for the contents of a human–machine friendship. And while the previous conditions have not been formulated in the negative, they are not concerning the qualities of a human–machine friendship, but the machine itself. The same could be said about a human having to be awake or being able to communicate in order to be friends (a person on the side of the world with no means to communicate with me, for example, cannot be my friend the same way a spaceship cannot be a machine friend).

7.6.3   Non-Coerciveness

This condition is necessary to avoid that the condition of exploitation is being circumvented. Synthetic friends cannot be coercive, as they otherwise would not be friends. This does not mean that they should not be able to change or influence humans. We should encourage that in human–machine friendships, synthetic friends can influence human friends and vice versa. This appears like an unavoidable condition of being in any kind of social relationship with someone else, as the interactions with them will eventually affect our purposes. This may even go deeper, as we could learn from synthetic friends, readjust our goals, and grow as a person. We can easily imagine that conversations with a synthetic friend can compel us to take life-changing measures.

The idea that machines should in any way be allowed to coerce us into anything, however, must lead to problematic outcomes and thus, the non-coerciveness of machines should be a necessary condition.

7.6.4   Non-Exploitativeness

One main worry, also already mentioned in Chapter 1, is that machines with which we built intimate relationships are exploiting the trust and reliance extended to them. This concern points toward the often problematic business models of the companies creating these synthetic friends. A subscription-based use of a synthetic friend could in the end mean the termination of the relationship if the user cannot afford the subscription anymore. This cannot constitute a friendship even in the wider sense of the word, as the synthetic friend would hold the human friend hostage. It also evokes a specifically transactional nature between the machine and the humans, which we also rule out.

The dimension of such non-exploitativeness essentially requires the disconnect from any kind of corporate interest within the use of the machine. A machine that is built to extract money, data, or other resources from intimate relationships must mean that this is never an intimate relationship to begin with as there are always third-party interests in the background of the interactions.

This does not mean that all synthetic friends should be free of charge or only given out to those with pathological needs—there can be a sense of difference of some synthetic friends that have other skills or have been more refined in their production. However, any friendship in which the danger of exploitation is built-in, not by the voluntary yet impermissible actions of a human but by the covert ability of a machine created for such a purpose, cannot be considered a friendship.

7.6.5   Non-Deception

John Danaher names in his paper on "Robot betrayal" (Danaher, 2021) several types in which robots can be deceptive to humans, and deception has been a wider concern within the human–machine-friendship discourse as an argument against creating these kinds of machines in the first place. The ability for humans to create machines that are incapable of misleading is one that should be appreciated and pushed for. While this has been exploited as an idea in science fiction, as robots are often portrayed to be incapable of lying or understanding sarcasm, the basic idea necessary to avoid the otherwise "mild" consequences of benevolent deception to accumulate to create harmful deception.

While we can think of ways to create machines that can lie or mislead, this norm should remain a necessary condition for human–machine friendships. Machines that can deceive are not only a slippery slope toward machines that create full illusions. Any exception we could create, i.e., based on the argument that we also lie or mislead in good faith to shield the other person from harm (we can safely assume that for many friendships, veracity is not an absolute value (contrary to what Kant may suggest, Kant [1977]) reintroduces human–human standards. The fact that we can create a machine incapable of misleading is an important starting point for these considerations.

This also seems to be the worry that Nyholm is addressing that has been mentioned before: in his view, we cannot build a machine that is not deceptive in this regard, and thus we should refrain from building them altogether. Considering that we elevate "non-deception" to a necessary condition, we should agree. If we cannot construct a machine that cannot deceive, we cannot create human–machine friendships. However, as we discuss elsewhere, we should be confident that we can. Additionally, there are ways of creating distinctly synthetic friends that refrain from coming into suspicion of being deceptive by giving them non-human traits.

7.6.6   Non-Exclusiveness

Another concern for the ability of machines to be friends consists in their potential lack of awareness of the person's wider social context. As we will discuss in Chapter 8, we should expect machines to mesh with people's social context to avoid some problematic consequences. However, we should acknowledge that there is a condition for machines to be friends that goes in the same direction: a machine must not claim the person's time or attention exclusively. We can easily imagine the previously discussed concern of companies creating "engagement machines" emerge if this condition is absent. A company producing synthetic friends might aim to keep its customers happy through easy engagement, suggesting that one goal would be that the synthetic friend works its way ever to increase the occupied time of the human friend. This must not be permitted, as a human–machine friendship, in which the machine aims to increase its share of attention it gets from the human friend, must lead to incentives to take over a person's social life. We can draw similarities to human–human friendships, in which one friend obsessing over another may damage the friendship. If one person becomes too demanding, we would eventually argue that they cannot be friends with each other anymore (it is possible, however, that two human friends are very close and do not mind the mutual demand for exclusive attention). Acknowledging that someone has more than one friend should be at least a condition for entering a human–machine friendship.

7.6.7   Non-Criminal/Violent

Human–machine friendships must not be violent ones. Some friends see it as part of the duties they have for each other to commit crimes when necessary. Take, for example, the willingness to defend each other with violence if the other is attacked, or some friends' willingness to lie to authorities to cover for each other. Especially in contexts of "chosen family"-friendships we can see a certain deep-seated alliance between friends, taking on rights for each other that the law only extends to family members.

A human–machine friendship cannot exhibit these features. A synthetic friend that is asked to break the law or do some violence should be programmed to refuse, rendering human–machine friendships a violent-free relationship. The reasons to consider this a necessary condition are clear from the potential consequences: a machine that can harm humans on purpose will harm humans eventually, and securely determining whether some violence is appropriate or not should not be left to the machine. Of course, this will not keep some humans from weaponizing their synthetic friends (as some people purposefully raise their pet dogs to be aggressive or seek friendships with other hateful people), but the intended quality of these machines must be to remain fully non-violent.

This condition is also inspired by Asimov's laws of robotics (Asimov, 1950), as they also outlaw a machine being programmed to intentionally harm a human. As Asimov makes this a statement on the permissibility of any kind of machine, this a fortiori applies to social machines intended to befriend humans.

This condition will leave synthetic friends to be incapable of rescuing their human friends by harming another. This, then, is another fundamental difference between synthetic friends and human friends, as the latter can reasonably judge that causing some harm to someone else is worth helping a friend in (physical) need. This difference, however, is necessary to secure a safe future with social machines.

7.7   CONSTRUCTIVE SUGGESTIONS

With the acknowledgment that human–machine friendships need not be modeled or evaluated after human–human friendships (as we would not model human–animal friendships are human–human friendships, either), we can embrace the differences these friendships can exhibit, within the necessary conditions which we argued for just before.

In opposite to the necessary negative conditions, these constructive points are mere suggestions and can vary from personal preference, cultural contexts, and general social development and fashion. As long as the negative conditions are being fulfilled, we can acknowledge that the machines in question can become synthetic friends to humans. We should, however, make some positive suggestions, as some elements that can make synthetic friends uniquely dispositioned to have an impact on our lives and, potentially, our relationships with other humans.

Simultaneously the discussion surrounding these constructive suggestions also concerned with possible more necessary conditions for synthetic friends. We thus not only justify these suggestions as helpful ones to aim for in synthetic friends but also justify why these suggestions are mere suggestions and not necessary conditions.

All these suggestions have an ethical dimension to them just like the necessary conditions—as we are concerned with how we should create synthetic friends, we are reflecting on the benefits these entities can have for us, and what we might owe them in return.

7.7.1 An Understanding of Human Emotion and Suffering

A not necessary condition for machines that can be elevated to synthetic friends, but still an important feature to have is a certain sensibility for the human conditions, especially the ability to suffer. In the following, we first distinguish between empathy and sympathy and second discuss whether, and if so, how a machine friend should be able to suffer.

7.7.2 Empathy

Empathy plays a crucial role in human relationships, especially in close relationships like friendships. Thus, it seems reasonable to require machines to be able to also have empathy in order for them to enter human–machine friendships. This requirement clearly has some appeal and has been subject to a diverse range of inquiries on the possibility and quality of empathy within machines, from recreating human-like empathy to proposing certain kinds of artificial empathy. Misselhorn (2021, 43ff.) identifies three elements of empathy and argues consequently that machines will not be able to be genuinely empathetic with us.

Her three elements are congruency, asymmetry, and a foreign consciousness. Congruency is required to recreate the same feeling in oneself, asymmetry for realizing that this is caused by somebody else's situation (i.e., the empathetic feeling is more appropriate to the other person's situation than my own), as well as another mind (Fremdbewusstsein) for knowing that it is someone else's state of mind that is replicated in oneself. One cannot be empathetic if they are representing a different feeling from the person they are empathetic with (incongruency), if they have reason to feel the same feelings from their own perspective (symmetry), or if the feelings are not represented as caused by someone else's feelings. Yet, as Misselhorn claims these three conditions to be both necessary and sufficient, once we do fulfill these criteria, we are empathetic. Thus, one question we can ask here is whether empathy with these criteria should be applied to synthetic friends.

The challenge for the question of empathy in human–machine friendships is, thus, the following: first, one has to answer whether some form of empathy is necessary for friendships to occur, and second, one has to answer whether machines have the theoretical capacity in the foreseeable future to be empathetic.

Some authors have answered this first question positively and assume that empathy is a necessary condition for friendship, as it is fundamental to understand human relationality from the perspective of emotions (Helm, 2022). This might be true. It seems that some philosophers have spoken against such a condition or can be interpreted to speak against it, like Danaher or Ryland, as they reject the condition of knowing other minds for friendship, either because it is not a condition at all (Ryland) or because we merely require reliable behavior for such a judgment (Danaher). Others, such as Nyholm (2020, 152) and Misselhorn (2021, 48) have stressed the importance of knowing that the other mind is indeed feeling the same or at least can feel the same.

If a machine that we intend to befriend does not have at least a behavioral response to a variety of human emotions, we should worry about the quality of that friendship. Thus, it does not seem too unreasonable to expect machines to be able to be equipped with some sensibilities toward human emotions.

One argument against synthetic friends at large, then, is the following: if compassion or a shared understanding of the phenomenal content of a person is albeit not a necessary condition then at least a very important one for any good friendship, we require machines that have similar phenomena.

Danaher has refuted this argument before by pointing out that, technically, we can never know whether another person is actually empathetic. Not only as a problem of other minds but as a problem of other's empathies. Just because someone expresses their empathy does not mean that they feel such empathy in the first place. And arguably, the comforting feeling of having empathy expressed toward us can be a performative one that gains appeal through reliable performances.

Thus, whether a machine expresses empathy should not decide on its ability to be a friend, although we do not deny or endorse that a machine that indeed can be empathetic is worth creating and studying.

7.7.3   Empathy vs. Sympathy

Misselhorn, as almost a side note in her book on artificial emotional intelligence and empathy, states that she does not believe that machines can fulfill her empathy criteria but what machines have become good at is expressing sympathy. Sympathy is understood here in the modern sense (in opposite to the sense in which it was used in the writings of the Scottish Enlightenment, as they meant it to mean mostly the same as empathy today). Sympathy is achieved if someone is encouragingly and with a goodwill is feeling with someone, or at least communicates these feelings toward someone. Thus, we can console someone or encourage someone, or hope for someone's success without having to feel empathetic but rather sympathetic. This limits the necessity to understand the other person's mental processes to our ability to judge their mental processes. Further, while empathy might be best described as a mental process with normative consequences, we can understand sympathy to be a primarily normative stance with some mental requirements. Without the ability to represent someone's feelings in ourselves, we cannot be empathetic.

Suppose we lose our ability to feel emotions, maybe due to severe depression or some brain dysfunction. If we are then told that someone is hurting or suffering, we still can be sympathetic with them, though we would not, according to Misselhorn's definition, be able to empathize. By expressing our wishes for them to feel better, or by offering help to make it so. Sympathy, thus, is a normative stance one takes toward another.

This contradicts Misselhorn's (and others's, like Fehige's [2004]) argument that empathy is the necessary condition for moral judgments, as the personal feeling of pain that we observe in others is the fundamental motivator to decrease suffering in the world.

Thus, from the perspective of empathy, anyone incapable of being empathetic, cannot be a moral judge. This is false. As we have seen, the knowledge about other people's suffering, rather than the representation of such suffering, is the minimum requirement for moral judgments. Thus, goodwill, the sympathy for someone else's success and well-being is the condition for such judgments rather than empathy. Additionally, even if we agreed that empathy is the beginning of moral judgments, we can transfer this knowledge to those that cannot feel empathy. The moral community may be built on empathy but is sustained by sympathy.

7.7.4   Sympathetic Synthetic Friends

Now, having seen that empathy might not be necessary for either being a moral agent or being a friend, we can suggest that a synthetic friend might be able to fulfill this condition, even if we do not judge it to be necessary nor do we have to agree with Misselhorn about machines being incapable of being empathetic. This question simply does not arise if we can create sympathetic synthetic machines. Misselhorn even goes so far as to acknowledge that we can create sympathetic machines (Misselhorn, 2021, 60).

From human–machine friendships, considering again their fundamental difference from human–human friendships, we can encourage the presence of sympathy rather than empathy. Even in emotionally challenging moments, we do not have to require that the friend we confide ourselves to has to be empathetic. They might "merely" say things to console, to encourage, to uplift, to rebuild, and to calm, but not to express that they know the feeling or can otherwise emotionally relate. We can also be very well aware of these limits in those friendships without questioning the ability to be friends in the first place. A synthetic friend might not be the friend we seek when we must be understood in our pain, for example in grief or when falling in love. However, for other emotional purposes, we do not have to assume that empathy is of the utmost importance. Knowing that we are being respected and that others acknowledge our success, our hard work, or our struggles, can be enough to feel heard, and maybe should be enough for an entity that we have previously excused from having to feel empathy.

The second point, however, is that synthetic friends may express their empathy in these non-feeling ways and could yet be comforting. Knowing culturally-specific and context-appropriate gestures and phrases to react to someone's stress or struggle is possible to the degree that it helps people. Whether we will ever expect synthetic friends to feel the same way we feel is questionable; we would also not complain about our pets for not relating to our emotions correctly at all times. If a synthetic friend fails at reacting appropriately, this might be awkward, but people do these things all the same; thus, there is harm to be expected from inappropriate attempts at, e.g., consoling a grieving person, although we may simply brush it aside as we would be a tail-wiggling dog who is interrupting a moment of deep sadness.

7.7.5   Suffering

If machines need not be able to feel empathy as a relatable feature but only sympathy, the question emerges whether they should be made to feel pain or suffering at all. Similar to the argument from empathy, those rejecting human–machine friendships often do so on the basis of machines missing the ability to suffer and thus missing out on some of the more fundamental features of life. And without such fundamental insight, the difference between humans and machines is so large that we cannot reasonably expect to be friends with them. This argument might even be stronger than the one from empathy. Empathy requires the ability to correctly read and represent someone else's mental state in oneself. The ability to suffer, being a condition for this, is thus more fundamental to relatability than empathy. We can understand someone else's misery without feeling empathetic if we can simply understand what it means to suffer. A machine, confronted with a person suffering from pain, trauma, anxiety, etc., will not grasp what that can mean to the person. Thus, their reaction will be potentially inappropriate as they cannot understand the state of mind a person is going through.

Further, in the interactions of humans and machines, lacking the ability to suffer (and us knowing that machines lack such ability) might trigger abuse from the human side. We see this in the discriminatory language of men toward female-gendered personal assistants (UNESCO, 2019) and in the treatment of machines by children when told that the machine cannot feel pain. Since the machine cannot suffer, there is no harm done. However, we should still want people to treat their synthetic friends with respect.

Here, a moral code may be helpful to create machines that, while they do not suffer, suggest equipping the machine with the ability to give an impression of being hurt. As part of moral education (or rather, to avoid moral deterioration), we should allow machines to intervene on their own behalf even if they do not suffer from the fact that they are treated to ensure that people do not get used to treating others a certain way. Synthetic friends must not become social punching bags that can be used to fling all the curse words at if someone is frustrated. Thus, there ought to be some strong limits on how you can treat synthetic friends until they either quit or retreat.

7.7.6   Playing to the Immortality

One of the main differences between machines and humans, or even between machines and any living organism, is the lack of mortality of the former. Should we ever be able to create consciousness machines comparable to human level (recall that we stated in Chapter 1 that we do not have to for this investigation to take off), we can be certain that a machine's view on existence—both it's own and one of the mortal beings—will be fundamentally different from our view.

This difference, I believe, can and should be utilized in human–machine friendships. We can name several reasons for why this should be done. First, we cannot mask a synthetic friend's fundamentally different constitution without deceiving the human in that friendship. Even with non-conscious machines, we are usually well aware that the machine will not age and grow with us in the same way. Pretending that machines have some grip on life as if their own existence was biologically limited would be ridiculous and inappropriate. Machines, as long as we do not make them age or die, do not age or die in this traditional sense. Now, we could easily create something that has a limited functioning time or create a "kill switch" that makes the machine shut down after a certain time. One could even argue that considering the contingencies of life, our other relationships with mortal entities (pets, other human friends), and our own mortality would suggest that we indeed want machines to be able to die. Sociality could, in this argument, require mortality.

Machine death would, however, also lead to some of the worries we can have with the underlying narcissism of tailor-making friends for us (cf. Sect. 7.3.3). If a machine is supposed to be able to die, it is most likely not supposed to die before the human—as that would be cruel and unnecessary harm. Thus, a synthetic friend is either supposed to die with us or after us. This would make them not only completely dependent on us as their only reason for their continued existence, but would also subject them to a status of social subordination. As mere attachments to our lives, their existence would end when ours end. The inevitable two-class social structure emerging from this is a reason to reject this point. Machines should not be slaves.

For a positive spin on this worry, we could argue that synthetic friends are given a reason to actually care for our well-being if their existence is connected to ours. However, this—again—creates something completely dependent on our existence and thus is more reminiscent of making machines our slaves and subordinates. Lastly, we can recall the argument of whether we should create machines that can suffer. We stated there that it seems unnecessary for synthetic friends, at least, to be able to suffer, as we noted that instead of empathy we should aim for sympathy. Creating machines that can suffer then is merely intended to somewhat improve human–machine friendships (if, for example, we wish to relate our suffering to machines as well). We concluded this to be an unjustifiable reason to create more suffering in the world, especially since we can always relate our suffering to other humans. Synthetic friends, from both a consequential calculus as well as from a deontological imperative to not create suffering, should not be made able to suffer.

This argument extends to the question of machine death and mortality in the same way. Independent of what we can do with immortal machines, how they will react to their own mortality, or our intention to switch them off, creating something that dies seems unjustifiable if we have the option to create the same thing without it having to be mortal. Machines will crash, or break, and be broken beyond repair, but creating them to eventually die only adds death to the world.

7.7.7   Secret Keeping, Memory Keeping

Another good use for synthetic friends is their lack of need to spread secrets. The urge to tell other people secrets and receive secrets from them has been widely researched about how we build relationships with each other. As an extension and test of each other's trust, as a way to build an exclusive common ground between two friends, and as a way to vent and release social and psychological pressures (Broeders, 2016), we tell each other secrets. However, often enough, telling each other secrets comes at the cost of those secrets coming out or us having to worry about having shared too much with not trustworthy people in a moment of weakness. The uncovering of a secret can come at great embarrassment and other harms, both to the person that told the secret as well as the person the secret was about. For all those purposes in which secrets are shared merely to gossip (if the disclosed information is about someone else) or to release some psychological tension (if the disclosed information is about oneself), a synthetic friend that gives some conversational feedback could be a very helpful social addition.

A synthetic friend could help us acknowledge our secrets without having to fear to be exposed, in a way that only a diary could. It is not a surprise that diaries and journals are often addressed as recipients of secrets, rather than mere notebooks in which we simply list our innermost truths we are not yet prepared to share. Thus, while not a necessary condition for human–machine friendship, the ability for synthetic friends being our confessional friends and cul-de-sacs of gossip should count as a constructive, justificatory point for creating these machines in the first place.

The opposite, however, is also true. As a sympathetic onlooker in our life, a synthetic friend can also record the most significant parts of our experiences, either with each other or with others. It can thus create memory profiles, assist us in memorizing important or nostalgic moments, can trigger these memories as a way to bring us to a calm mental place, and create personalized clusters of memories that help us make sense of our lives.

Such a feature is often portrayed as a step toward a dystopian future in which every ever so embarrassing detail will be recorded and remembered, undermining the quality of our memory and of creating a narrative of our own life that ignores some memories that might not fit our requirements for coherence. There is a substantial risk of the omnipresence of this kind of technology in our current way of life, and without blindly endorsing a future in which vital mental processes like memory, self-understanding, and other identity-building exercises are delegated to technology, we should carefully reflect on the potential fallout of such technology.

However, on the one side, we could argue that we are already there: smartphones already assemble little memory movies in which an algorithm analyzes the pictures and interprets these memories on its own. We even voluntarily store evidence of important life events (images, videos, text messages, etc.) so as to remain able to access them. Whether an autonomous, social machine like a synthetic friend would change our perception of the memories stored with them is an open question, but I am doubtful it will disrupt our own narrative of life too much. Rather, it will enhance it and guarantee that some of the most cherished memories will be available to us at will. On the other hand, this might be a genuine improvement over some of life's restrictions, especially in the latter days of someone's life. A synthetic friend, which has grown close to a human and has been storing memories (and thus extending that person's memory), can have an active role in retaining and recreating those memories that would otherwise be lost. The resources a synthetic friend could provide to anyone who would like to remember their own life better are incomparable to what has been available thus far.

Certainly, this technology should be carefully weighed and might ultimately be rejected. Plausibly, the human mind must forget in order to survive, which means we should not have access to digitally stored memories through synthetic friends. We can imagine that the consequences of having access to our stored life stories would be obsessing over past mistakes and the ensuing regret, the risk of getting lost in times when one was happier by not moving on from them, or the lack of ambiguity necessary to create our life narrative. Arguably, the human ability to forget is one we should endorse rather than fight. Yet, synthetic friends could be a reservoir of those memories that the human friend considers that must not be forgotten. Ultimately, the potential for synthetic friends' unique disposition as a machine, and the resulting unique features of a human–machine friendship can reasonably show in the differences of how those two friends remember things. While human memory is connected to a variety of elements—senses, emotions, moods, thoughts, and narratives—next to the actual events remembered, a machine is limited to storing the events that unfolded. While we have seen that merely providing this information at every turn of a person's life is likely unwanted, we should remain open to the opportunity to help people maintain and improve their sense of self and their own narrative of life. A synthetic friend provides the needed information for that.

7.8   Social Markers and Anthropomorphization

One last critical point, both in concern and in opportunity is the question of the degree of anthropomorphization. As we have discussed elsewhere (Kempt, 2020, Chapter 5.1.3), anthropomorphizing machines is not an ethically or politically neutral choice and thus should be addressed in human–machine friendship concerns. Depending on what features of human beings are being recreated in machines, the consequences of such decisions for the perception of the machine, and the perception of the imitated humans, can be substantial.

These consequences are of both ethical as well as political concern. Without knowing how widespread some technological device will become, its "politics of artifacts" (Winner, 1980) will remain opaque or otherwise difficult to anticipate, with especially relevant consequences for machines taking over human tasks (Haraway, 1985). Movements of "inclusive design" have been taking up the task to anticipate certain inaccessible design choices for specific groups that might only take effect once this technology is widespread and applied (which would ameliorate any discriminatory features of that technology a difficult enterprise). Thus, the basic idea is that every technology, being assumptive about its potential users, its conditions of production, and its intended use, is also always a political artifact. Reflecting and justifying the politics of these artifacts is thus part of a normative analysis of technology. If we find these politics are indefensible, we might judge the technology to be indefensible altogether. Take, for example, anti-homelessness architecture in cities: benches on which one cannot lay down and rain-protected empty spaces with spikes to avoid people gathering there, to only name two (De Fine Licht, 2017). These are artifacts with politics built into them, as they are specifically designed to discourage humans from using them for their needs other than the narrowly defined ones. Taking the ability of an already marginalized group of people to sleep is cruel and harmful, as the very intention is to keep the homeless from using those public spaces and facilities. Thus, the technology should be judged to be indefensible as well, even if they are merely spikes on the ground or an oddly shaped (potentially even aesthetically pleasing) bench.

Why should we bother discussing inclusive design in the context of anthropomorphization and more specifically human–machine friendships? It seems that inclusive design approaches face a dilemma here which we should discuss before turning our attention toward whether a solution to this dilemma is workable for synthetic friends.

7.8.1   Anthropomorphism, Anthropomorphization

Before we can discuss this dilemma, however, some clarifying notes on what we understand with anthropomorphization. Anthropomorphism, generally speaking, is the process of ascribing human features onto non-human entities. However, we should make a distinction between anthropomorphism and anthropomorphization. The latter is the process of projecting human features onto non-human entities, while the former can be understood as an active process to create something that can be anthropomorphized (cf. Jones's use of "projective anthropomorphism" for what we call "anthropomorphization", Jones, 2021).

This distinction is relevant because it shows the two dimensions at work here that leads to the anthropomorphization of machines: the willingness of users to project human features onto a machine and thus suspending their disbelief about it, and the creation of a projection surface that is enabling, if not provoking those projections. This is often a design choice to create either an illusion of the machine being more human-like (and thus generally more capable than it actually is), or to simply imitate humans to replace them in certain areas of the labor market. In order to assess the process behind anthropomorphizing technology, we should be explicit about the double-edged process: social machines are built to be anthropomorphized by being equipped with human-specific features, and users are willing to take these features as almost real.

Take, for example, human willingness to anthropomorphize pet behavior. Some humans talk about their dogs with the same terminology (and expectation for social integration) as they would use for describing humans, based on some behaviors that are suggestive of human behavior but otherwise unfounded to be characterized in the same way. Social machines, exhibiting similar behaviors are similarly anthropomorphized, while such projection is similarly unfounded on the actual mechanics of the machine.

7.8.2   The Pro of Anthropomorphization

With this distinction in mind, we can turn toward some of the positive features of such anthropomorphism and anthropomorphization. The main argument for anthropomorphization is that it makes social machines more accessible as it reduces the need for human users to align their expectations or interactive norms toward that machine. If I can ask a machine to do something for me the same way I would ask another human, I do not have to learn anything new in terms of interactive norms. Further, if a machine follows the same conversational rules as I, the chances of being unduly influenced by the machine are smaller as the machines had to follow established rules of interactions.

It further diminishes frustration if a malfunction or some other problem emerges as it can use conversational tools to appeal to the understanding of the user. The fact that a personal assistant apologizes if a task cannot be completed, or even uses sentences in the first person singular is anthropomorphization techniques to make the machine more relatable and thus motivate some understanding and tolerance toward the machine that we would grant to other people, too. The more anthropomorphic a machine, thus, the more accessible it should be for most users as there is less alignment to be had. Suppose that this can be done without leading to deception, this would mean that accessibility is increased through the ease with which we can align and incorporate these social machines into our lives.

From an inclusive design perspective, then, we should demand anthropomorphization for these technologies to increase the ease with which people can use this technology. As we can specify these requirements of social machines for synthetic friends, we could argue that synthetic friends should have anthropomorphic features, e.g., human-style conversational habits, that make the technology more intuitive to operate and thus easier to relate to. The more foreign or artificial these technologies appear, the less people will be able to use them properly.

There might be an immediately intuitive point to support and reinforce this resistance to anthropomorphization. Maybe we should want that people are challenged by the features of these social machines and only those that exhibit a certain tenacity and willingness to align their social expectations. However, as we have seen from the core argument (Sect. 7.2), once we acknowledge the possibility of human–machine friendships, we also have to acknowledge that we can recommend these to others. Thus, we cannot advocate for intentionally designing synthetic friends in a way that limits their appeal to those who are more committed or more tech-savvy or simply in more pressing need for such a relationship. Further, as we will discuss in the next chapter, we should want those machines to be able to integrate socially, which would require some relatability for those who do not wish to relate to fully alien machines.

It seems, then, that we should create relatable machines using anthropomorphization techniques as these make the technology more accessible, more integratable, and more inclusive.

On the other hand, if one aims to create technology that is not only inclusive to all humans through increased accessibility but *about* humans, the politics of the artifact become essentially the design-subject themselves. This is the point made earlier about the politics and ethics of imitating humans. Not only must creators of anthropomorphized technology acknowledge that their product will cater to a certain group of consumers but they also have to acknowledge that the means to reach these groups is to rely on prevalent social stereotypes and typical human behavior. These stereotypes, the behavioral patterns, and the norms behind those behaviors are grown in hierarchical and potentially oppressive power structures. Clearly, not all are born from these structures, as some behaviors are ethically justified generally accepted ways, e.g., greeting someone and engaging in small talk.

However, finding the level of anthropomorphization that is not reproducing problematic stereotypes of different groups of human beings is a highly problematic affair, especially when it comes to otherwise innocuous concerns like norms of politeness (Kumar et al., 2021) or rules for what counts as small talk, what is appropriate to ask whom in such small talk conversations, and how to react to certain answers. The burden of proof lies with those claiming that anthropomorphization can be done while not engaging in these questions of appropriateness.

From this insight, we can distinguish further between "mere anthropomorphization" and "social anthropomorphization", i.e., an anthropomorphization based on social markers. Mere anthropomorphization, in this sense, is the use of human features in technologies that are not necessary for the machine to work. A machine using a natural language, for example, English, is thus not an anthropomorphization because otherwise the machine would not work. If, however, the machine is using English in a way that would suggest that there are other human features represented, we should call this anthropomorphization. Social anthropomorphization, on the other hand, is achieved when these anthropomorphic features are not only suggesting humanness but tell you something about the character that is displayed with these features. Equipping a chatbot to be able to state "I am doing well today", in this definition, is mere anthropomorphizing, as we do not learn more about the chatbot except for pretending to do well (while it is in fact not able to be well or unwell at all). If we give the chatbot a female-gendered name, or make it use more contextualized language (as in stating "I am doing well, sweetie"), the social anthropomorphization becomes obvious.

The main concern in anthropomorphization, then, emerges when this question extends to social markers, i.e., humanizing features that are intended to also imply certain social features of the machine. These are often used to increase the relatability of a machine that is intended to engage more thoroughly with humans in their surroundings. Social anthropormorphization, then, should be investigated more thoroughly in regards to its effects on human–machine relationships.

7.8.3   Anthropomorphism in Human–Machine Friendships

The question in this context is whether we should allow synthetic friends to be socially anthropomorphizable, or whether instead there should be a necessary condition of non-anthropomorphism in which adding human social markers are shown to be impermissible. "Mere anthropomorphization", on this level of social machines, seems highly unavoidable as we otherwise would not be able to relate to machines at all.

Non-social-anthropomorphizing, one could argue, should be a necessary condition for human–machine friendships, as it otherwise enables (and foreseeably guarantees) for the recreation, reiteration, and maintaining harmful stereotypes or other misleading social cues. A relevant feature for friendships and any other more intimate social relationship as well and thus a good example of how these social anthropomorphizations work is gender.

Gender, as a social marker, is important to many people's behavior around others (even if it should not matter, it is in fact a strong predictor for different behaviors). It is thus not surprising, then that many social machines have been gendered in one way or another and thus, in our terminology, have been "socially anthropomorphized".

7.8.4   Gender

To judge whether we should allow synthetic friends to be gendered or not we should specify that it means to be "gendered". Clearly, some gendered technologies come to mind to demonstrate what it could mean: personal digital assistants that use a typically female or male sounding voice, gendered names for certain social machines (Amazon's Alexa, Microsoft's

Cortana, Google's Meena, etc.), the choice of avatars and images to represent otherwise unembodied machines (chatbots or personal assistants, like the chatbot Kuki (Pandorabots, 2022) and the like. These measures are intended to make it easier for consumers to relate to these machines and, as a result, use it more often, more confidently, and with more engagement. It seems easier for consumers to navigate social interactions with a gendered machine very much because of the anthropomorphization, however problematic this process may be. However, as pointed out above, for the interpretation of a machine as gendered we both need anthropomorphism and anthropomorphization: without both the provision of a projection surface, through female names, imitating "female" ways of speaking, and a female-presenting avatar on the one, and the willingness to suspend one's disbelief that these projection surfaces are, indeed, merely projection surfaces, the perception of a machine as female would not succeed.

The special concern with socially anthropomorphic features is that these usually spread: if a machine is not only gendered by name and voice, it is often used assumed that there are also other features present or gendered norms of behavior apply. Social markers imply social status. Thus, when we discuss whether a machine should be gendered as in "thoroughly represent the (stereotypical) features of a gender", i.e., should minimize the need for consumers to project, the answer is a clear no. Machines should not represent genders in this way. The politics of this artifact are not defensible.

However, if we understand the question of assigning social markers to mean whether a machine should exhibit any kind of typically gendered feature, it is less clear and potentially irrelevant. It is less clear because some currently gendered features may not be associated with a specific gender performance in the future, or because without a full gendered performance these short imitations are not leading toward full gender association while still providing some increased relatability for those using machines.

And it is potentially irrelevant because first, by creating the need to project and assume a lot about the machine, we may not associate the small instance with an overall gender and thus harmful stereotypes may be less powerful or prevalent. Take, for example, a female-gendered name without any other gender marker of a woman—a robotic voice of ambiguous tonality, no avatar. We would not expect anyone to assume a woman to be "behind" that machine or to be represented by that machine. However, the more elements are confirming the gender—"identity" of the machine, the more likely it is that someone assumes there to be ever more gender markers. Thus, an occasional reference to something that could be interpreted as a gender does not mean it will lead to the emergence of harmful stereotypes being recreated and projected onto the machine.

And second, whether we create a genderless social machine and insist on it lacking a gender, it will not stop people from inventing a gender story for the machine if they so choose. The projection of gender is a bigger concern here than the decontextualized reproduction of it for increased relatability, even if the reasons to project and anthropomorphize are rather slim. However, there is reason to be optimistic about the ability of humans to not require too many social markers to build a relationship with machines. Synthetic friends might end up taking firm places as simply genderless entities in our lives.

There are, however, undeniable advantages of using social anthropomorphic social markers: we should consider the relatability of social markers, even if merely performative, can be for enabling increased interactions between machines and humans. We usually require knowledge of the other person to relate to them, from names or the tone of their voice to more complex information about them. Without giving machines something of that kind would render them largely unrelatable. The current field of relatable social machines is only successful precisely because most of its examples (Kuki, Replika, Alexa) are playing into social marker stereotypes, especially gendered ones. For many embodied ones, like sex robots, these stereotypes are almost mandatory (Devlin, 2018, 136). It stands to assume that for most startups and small companies, to get a foot off the ground and build a client base, as well as for large corporations having to produce artifacts with non-challenging politics built into them, resorting to gender is an easy, effective, and thus lasting heuristic to attract customers.

However, we can ask whether the actual relatability should be of consideration here. For the success of those machines it might be necessary for them to achieve a higher level of anthropomorphism and anthropormophization, as otherwise people simply cannot feel comfortable with these machines. Yet, the question which social marker to use, and to what degree, can make an impact on how we view the discourse on anthropomorphism. As stated before (Kempt, 2020, 113) nobody would suggest to create a machine that recreates racial markers rather than gender markers. Giving social machines attributes that suggest that they are black, in this sense, is equally nonsense as giving them features that suggest they are female. The fact that gender is a better predictor of behavior should not decide on whether we should endorse and push for those. There is, when it comes to creating social machines, simply no good reason besides increased engagement that suggest favoring gender over other social markers. And considering that appears important to many people as such a marker for relatability, we should ask whether the protocols and norms of behavior associated with gender are not precisely those that are grown in unfavorable, if not straight-up oppressive power structures (Haraway, 1985; Tollon & Naidoo, 2021).

We can feel comfortable to reject any bigger social marker, i.e., one that suggests a human personality, rather than simply "human features", to be of relevance for human–machine relationships even if those would increase engagement and potentially the number of people who would be willing to enter those friendships. The danger that gendered or otherwise socially marked synthetic friends would outweigh the increased engagement. This does not mean that we cannot imagine artificial personalities with human features ("mere anthropomorphization"), i.e., the ability to detect and use sarcasm, humor, or even sass.

7.8.5   The Issue with Deception

Thus far, weighing reasons for and against anthropomorphization through social markers has ignored the issue of deception and only considered relatability. As we found that synthetic friends must not deceive as a necessary condition for human–machine friendships, the line between anthropomorphism and anthropomorphization grows more relevant. If a machine is merely suggestive of certain markers but remains transparent about its machine-ness, it seems difficult to speak of deception of the user, even if the said user finds themselves treating the machines as a human. Humans tend to anthropomorphize rather quickly, and it would be a problem of applying the problematic dimension of anthropomorphism if we spoke of deception if a machine is being anthropomorphized (take, for example, the current developments often discussed in literature about people anthropomorphizing bomb detection [Carpenter, 2015]) or cleaning robots. We can be mistaken about a machine's capacity without the machine deceiving us in interaction or in presentation.

However, we should be aware that at some degree of social markers, they will eventually lead to justified confusion and, thus, deception about the nature and capacities of the interactant. The difficulty then is to find when somebody is "justifiedly confused" as a measure for deception.

We should expect a social negotiation process of what can be expected to lead to more specific answers for this issue. We can imagine the requirement for machines to self-ID as machines to break with an unintentional suspension of disbelief of a user, or to propose certain industry standards of what kind of anthropomorphic features are allowed and which are not. The decision when something should specifically count as deceptive, however, cannot be made by one person, or even by one discipline but requires interdisciplinary work on digital literacy, technological possibilities, and normative limits.

First, the main concern appreciates the risk of deception. An anthropomorphic social machine that behaves like a human person would usually suggests to also be limited to the same physiological boundaries we face. Danaher (2021) however correctly points out that this need not be the case, as a robot with front-facing eyes might as well have more cameras installed around their body, thus being able to see more or from different perspectives than humans ever would consider. Similar applies to the access to information on the internet: a human being usually is limited to their knowledge at that moment, while a social machine has access to most information on the internet in an instant. Anthropomorphism can mask these imbalances in a deceptive way (Chapter 5): the more human-like a machine, the easier it is for human interactants to suspend their disbelief that they are in fact talking to a machine. They might presuppose certain abilities that the machine should have or limits that the machine does not have. The harm that can follow from these vulnerabilities is an often problematized issue in discussions surrounding anthropomorphism.

7.9   Conclusion

This chapter aimed at providing a strong argument in favor of human–machine friendships. Several such arguments have been proposed before, usually aiming at demonstrating how friendships are not something fundamentally human as to exclude machines from joining the circle of friends. From Danaher's behaviorist adjustments to Ryland's degree approach, the attempts thus far have been aimed at including machines in the given philosophical theories of friendship. In this investigation, we provided an argument that leaves this line of thought to propose an endorsement of the fundamental differences in human–human and human–machine friendships. Instead of struggling to prove how machines can join the circle of friends, we acknowledged that the term "friendship" is already rich in variation on conceptual and normative conditions and rather representative of a family resemblance. With this acknowledgment, however, comes a new task: characterizing the conditions that machines should exhibit so we can call anything between machines and humans a friendship. For this, we proposed negative conditions that must not be violated, and constructive suggestions that would be conducive for a deeper relationship between humans and machines in this specific human–machine friendship approach. We proposed to call machines that fulfill the negative conditions "synthetic friends", to distinguish them from other, "merely artificial" agents.

The negative conditions, paired with the precondition of dealing with relatable technology in the first place, consist in the machine being non-coercive, non-deceptive, non-violent, and non-exclusive. While this limits the machines to be friends of a certain quality, we can see that as a confirmation that human–machine friendships are not a limited version of human–human friendships, but something genuinely new.

These conditions are necessary to avoid these fundamental differences between machines and humans to create lopsided, unfair, harmful, or isolating relationships with these machines. The constructive suggestions, on the other hand, are born from the insight that once a friendship is possible between humans and machines, and this friendship is something different from human–human friendship, then we might play to the strengths of such machines. Having a friend that does not suffer but can display sympathy for any situation can be an asset in our struggles with our own sorrows; a friend that can keep secrets without judging and retrieve long-lost memories or details of our past can help us understand our own life better and assist in creating a life narrative that can give orientation for our future; a friend that does not age or die can guide our life as a constant where there are often variables.

Therefore, synthetic friends are good candidates for human–machine friendships, as they preserve humans from harm within these relationships, enable all sorts of positive effects for their humans, and, depending on the specifics of their creation, might even add unique features and value to human lives. In the next step, we will explore the ability and our need for socially integrating these relationships as we find that a concentration of humans on synthetic friends alone may lead to problematic consequences altogether. Thus, we need to contend with the question of how to socially integrate synthetic friends.

REFERENCES

Asimov, I. (1950). *I, Robot.* Gnome.

Branigan, H., Pickering, M. J., Pearson, J., & McLean, J. F. (2010). Linguistic alignment between people and computers. *Journal of Pragmatics, 42*(9), 2355–2368. https://doi.org/10.1016/j.pragma.2009.12.012

Broeders, D. (2016). The secret in the information society. *Philosophy & Technology, 29*, 293–305. https://doi.org/10.1007/s13347-016-0217-3

Carpenter, J. (2015). *Culture and human–robot interaction in militarized spaces: A war story.* Taylor & Francis.

Darling, K. (2017). 'Who's Johnny?' Anthropomorphic framing in human-robot interaction, integration, and policy. In P. Lin, G. Bekey, K. Abney, & R. Jenkins (Eds.), *Robot ethics 2.0.* Oxford University Press. http://dx.doi.org/10.2139/ssrn.2588669

Darling, K. (2021). *The new breed: What our history with animals reveals about our future with robots.* Henry Holt.

De Fine Licht, K. (2017). Hostile urban architecture: A critical discussion of the seemingly offensive art of keeping people away. Ettik i praksis. *Nordic Journal of Applied Ethics, 11*(2), 27–44.

Devlin, K. (2018). *Turned on: Science, sex and robots.* Bloomsbury.

Fehige, C. (2004). *Soll ich?* Reclam.

Floridi, L., & Chiriatti, M. (2020). GPT-3: Its nature, scope, limits, and consequences. *Minds and Machines, 30*, 681–694.

Giubilini, A., Savulescu, J. (2018). The artificial moral advisor: The "ideal observer" meets artificial intelligence. *Philosophy & Technology, 31*, 169–188. https://doi.org/10.1007/s13347-017-0285-z

Haraway, D. (1985). *A Cyborg manifesto: Science, technology, and socialist feminism in the 1980s.* Center for Social Research and Education.

Helm, B. (2022). *Friendship: Stanford encyclopedia of philosophy.* https://plato.stanford.edu/entries/friendship/. Last accessed 15 June 2022.

Jeske, D. (2019). *Friendship and social media: A philosophical exploration.* Routledge.

Johnson, D. G., & Verdicchio, M. (2018). Why robots should not be treated like animals. *Ethics and Information Technology, 20*, 291–301. https://doi.org/10.1007/s10676-018-9481-5

Jones, R. A. (2021). Projective anthropomorphism as a dialogue with ourselves. *International Journal of Social Robotics.* https://doi.org/10.1007/s12369-021-00793-7

Kant, I. (1977). Werke in zwölf Bänden. Band 8, Frankfurt am Main 1977. Erstdruck in: Berlinische Blätter, 1. Jg., 301–314.

Kempt, H. (2020). *Chatbots and the domestication of AI*. Springer International.

Kumar, S., Itzhak, E., Olatunji, S., Sarne-Fleischmann, V., Tractinsky, N., Nimrod, G., & Edan, Y. (2021). *Exploratory evaluation of politeness in human–robot interaction*. arXiv. https://arxiv.org/pdf/2103.08441.pdf. Last accessed 15 June 2022.

McKeever, N. (2022). Online dating and love robots. In A. Grahle, N. McKeever, & J. Saunders (Eds.), *Philosophy of love in the past, present, and future*. Routledge. https://doi.org/10.4324/9781003014331

Misselhorn, C. (2021). *Künstliche Intelligenz und Empathie. Vom Leben mit Emotions-erkennung, Sexrobotern und Co*. Reclam.

Nyholm, S. (2020). *Humans and robots: Ethics, agency, and anthropomorphism*. Rowman and Littlefield.

Pandorabots. (2022). *Kuki AI*. https://www.kuki.ai/iconiq-ethics. Last accessed 15 June 2022.

Tollon, F., & Naidoo, K. (2021). On and beyond artifacts in moral relations: Accounting for power and violence in Coeckelbergh's social relationism. *AI and Society*. https://doi.org/10.1007/s00146-021-01303-z

Turkle, S. (2011). *Alone together*. Basic Books.

UNESCO. (2019). *I'd blush if I could: Closing gender divides in digital skills through education*. https://unesdoc.unesco.org/ark:/48223/pf0000367416. page=1. Last accessed 15 June 2022.

Winner, L. (1980). Do artifacts have politics? *Daedalus, 109*(1), 121–136.

CHAPTER 8

Social Integration

Many philosophical approaches to friendship conceptualize friendships as a bond between two people who typically are not related to each other. Their bond is voluntary and mutually enriching. However, what is often overlooked is the fact that many friendships are not a discreet part of someone's life, or that the bond between two people is an isolated occurrence. Usually, friends are an embedded part of someone's life, with other friends knowing each other, or even being friends with each other as well. The friends' respective families are usually aware of the friendship, as well as others (work colleagues, neighbors). Some friends have entered quasi-familial relationships in the form of responsibility communities and co-habitation setups. Sometimes, the line between friendship and romantic relationship is getting blurred, when highly intimate physical contacts are "allowed" between friends. Friendships, then, are a much more varied, rich, integrated or discreet, open or isolated affair. The general characterization of friendships must miss some key elements that make specific friendships to the perceived unique connections we establish. Most philosophical endeavors, as we have seen before, concentrate on the one-on-one dimension of beings friends. For most purposes of exploring the nature of friendship, this is a sensible approach, as defining a friendship through the features it contains in the micro-interactions and the medium-ranged shared activities. However, we reflect on the

H. Kempt, *Synthetic Friends*,
https://doi.org/10.1007/978-3-031-13631-3_8

social consequences of this new kind of friendship if we wish to present a comprehensive proposal.

One of the positive suggestions we should make in a human–machine friendship proposal, then, is how the social integration of those friendships (not necessarily of those machines) can be successful. In Kempt (2020), we introduced the idea of robophobes and robophiles as an approximation of potential fault lines in society, with Margaret Archer developing a similar pair of concepts later on (Archer, 2021). Though revisiting the full theory behind these two concepts appears unnecessary, we should still shortly reintroduce the idea.

8.1   Robophobes and Robophiles

Not everyone has the same approach to technology. In fact, the variety of technology adoption (early vs. late adoption, intense vs. occasional use etc.) is great and growing. Usually, there is a generational divide at play, with older people being rather luddite in their adoption of new technology and younger generations being more open toward them. The reactions toward new technology is also varied. From enthusiastic queuing by fans of a technology brand to purchase the latest generation smartphone to a certain cultural pessimism about the "take over" of technology in our life, the assessment of the promises and perils of new technology is not new.

What is indeed new, though, is the fact that social machines will soon become part of this tension. With human–machine relationships becoming more viable or attractive to some due to the increasing sophistication of those machines, we can expect at least similar, if not stronger reactions of those dismaying this technology and those who endorse it. However, while we may simply take this as another instance of the unavoidable differences in attitudes toward technology among people, it will have more severe consequences worth considering. While adopting new technology by one person may invoke ridicule or lack of understanding by another, this usually only affects the personal choices of the technology user. With the rise of social machines, however, we may anticipate that such ridicule will eventually become hurtful: if I begin establishing human–machine friendships, and someone else rejects the technology of social machines altogether, it becomes hard to see how they will be able to recognize my relationship with that machine.

We can call these attitude-extremes robophilia and robophobia, as they should be understood as a spectrum or range of attitudes one can exhibit toward social machines, from enthusiasm to complete rejection. The rejection or enthusiasm for these machines is morally indifferent (as we stated before, nobody can be asked to develop relationships with these machines, as nobody can be asked to develop friendships with anyone in particular [or, potentially, at all with anyone]). It is merely a statement on the willingness of a person to engage with social machines. The middle of the spectrum would consist in a neutral attitude toward those social machines.

8.1.1   The Problem with Robophobes

However, it becomes problematic when those friendships mature to the ones that we integrate into social networks. As we have pointed out before, many friendships are inherently embedded into people's social life, they participate and contribute to a person's connections, which in turn increase the quality of life. However, what if we want to introduce a machine to someone who is explicitly a robophobe for his own social connections?

We also just stated that is it morally indifferent if anyone would consider being friends with a machine, leaving the position of robophobia generally morally neutral. Then, how can we expect those who strongly reject human–machine friendships to interact with machines in a friendly way, which seems unavoidable in the future? And what should we expect from those robophobes morally? It seems necessary that any comprehensive approach to human–machine friendships should provide an answer to these questions. Especially because the danger emerges that robophiles (again, another morally neutral position) will lead to social isolation from other humans, reinforcing the dependency on human–machine relationships, if they are being rejected by a robophobic environment.

We see developments of this kind already in current robophilic communities. These are groups of similar-spirited persons that have some form of relationship to a robot they claim to be socially significant. While these are mainly concerned with sexbots and their emotional bonds to these machines, they are serious about their relationships. They found networking groups, have meetups, and exchanged experiences about what it means to them to love a robot. These networks grow, in part, because others in their social environment do not understand their position.

We can expect increased social networking efforts to connect those willing to engage with social machines to create social bonds, with inner differentiation about the different kinds of relationships. The reasons for these networks may lie partly in the uniqueness of these relationships and the need to share experiences from these relationships with others who can understand some of the phenomenology going on in them.

However, we should not ignore the fact that these relationships are not only qualitatively unique and thus people not in them will have a hard time understanding the lived experiences of such early adopter robophiles, but also that it is virtually impossible to share any feelings for these machines without being declared mentally ill, weird, perverted, or really just being looked down on. This must lead to isolation from other human beings in their life, as the acceptance of their lifestyles and choices. As Devlin (2018, 140) points out, most of those already connected robophiles are not pathologically asocial, or incels, or otherwise psychologically limited. They are, in opposite, capable of organizing themselves their interests with others.

If we wish to avoid growing tensions between robophiles and robophobes, we should encourage tolerance. Fundamentally different philosophical perspectives on the question of what it means to be in a fulfilling social relationship will affect these groups and their ability to communicate about these issues. We have precedence for this, for example in form of pet owners and their wishes to be accommodated by their surroundings even if they interact with people who do not care for pets in the first place. In turn, however, pet owners do have to accept that allergies, anxieties, and some other life conditions preclude them from entering every space with their pet present. Thus, we can expect both robophiles and robophobes to make some concessions.

However, in the long term, we should aim to reduce Thea mount of robophobia. As we have stated before, machines must not be slaves and merely seen in subordinating relation to their "owner" as this can easily lead to highly problematic two-class, separated societies.

8.2   Corrosion and Atrophy of Sociality

John Danaher discusses an argument on the potentially bad outcomes of outsourcing human connections and, how in his view we might be more inclined to instead add machines to our friend groups as they can "help" in improving our human friendships. This argument also emerges with other authors, like Vallor (2015).

The worry about outsourcing otherwise purely human connections has become somewhat of a standard argument against human–machine friendships. We previously encountered similar arguments in Chapters 1 and 3. In Chapter 1, the concept of technosolutionism was introduced to discuss the concern one can have when answering the question "what problem are we solving with human–machine friendships?" It appeared that we might attempt in answering a social problem with a technological solution, and the relinquishing our duties and delegating them to some technological item. In Chapter 3, we discussed how the concept of friendship may include a certain ambiguity tolerance, and that our ability to withstand and even appreciate differences between humans are a sign of a friendship that allows for personal growth and interpersonal recognition. If we were only to encounter pleasing people, we would not grow as a person, and machines appear to be poised to provide exactly that kind of constant personal validation that might keep us at the same developmental level.

The worry here, however, is a more specific one about the effects of creating machines that are too socially successful without being socially integrative that they would enrich society. Usually, the argument goes something like this: If we build social machines that are easier to deal with, validate and confirm our wishes and preferences without us having to exhaustingly negotiate these wishes and preferences with other people, we might be inclined to seek the companionship and friendship with these machines. This seeking of connection to a machine must come at the expense of us seeking human connections, and we might be naturally inclined to seek places in which we are validated for the lowest personal price to pay. Thus, we might mid or long term prefer human–machine friendships over human–human friendships, for reasons of easy validation. The consequences of preferring machines over humans will make us more self-involved and generally worse people. Even more, our engagement with other humans may *corrode* and *atrophy*, because of the standards being set by machines. Whether this is a voluntary process of simply preferring the less stressful, more patient, less judgmental, more consistent company of a machine friend, or the inability to invest the patience needed for long-term interactions with human beings is unclear in this argument, but often presented together.

In the following, we will analyze the two different dimensions of this argument, as it appears worth investigating whether the voluntary replacement or the involuntary loss of social skills will be relevant for the development of social machines. It becomes noticeable, however, that there is an external option to avoid these issues that ought to be explored in more detail, reconnecting the debate about robophobia and robophilia with this issue.

8.2.1   Social Corrosion

The voluntariness thesis in this argument may be the less concerning of the two options for two reasons. First, if most people simply prefer to be friends with machines over humans and are aware of the implications and considerations, they make a choice to be respected. As we said before, it should count as morally indifferent whether someone prefers relationships with a machine or rejects them—we cannot reasonably require someone to prefer robots or humans over the other in this context. Thus, even if we should disagree with the outcome, generally speaking we only have prejudice to hold against a society that works well with only human–machine relationships.

Second, this is unlikely to happen in the first place. The quality of human–human friendship remains objectively the same, as an alternative to a social connection usually does not affect said connection. The fact that we might be able to enter human–machine friendships that are pleasant or especially insightful for ourselves, as the machines may have some features that we miss in people in general, does in itself not suggest that anyone would long term prefer machine friendship over human ones. Certainly, some replacements will occur, but so far any kind of social technology has merely improved our range of connections. If social machines can, then, improve our connections generally, why should we be concerned that our other connections to other human beings will corrode? Rather it stands to assume that our dearest connection will be deepened, and new deeper connections will be made possible. Corroding relationships with rather loose connections may lead to reinvigorated connections with our closest friends, or allows for frictionless interactions with our friends, allowing for these friendships to grow to previously impossible levels.

The assumption about corroding connections due to "pleasing machines" is understandable, as we can easily imagine a reduction in superficial connections with humans if we can have a similarly pleasing interaction with machines. However, considering that we are surrounded by pleasing media already and that there is a chance to create machines that are actually integrative into the social context, we should be optimistic about these machines not "luring" people away from society, or that we begin to view humans as unpleasing or insufficiently pleasing.

8.2.2   Social Atrophy

The inability thesis on the other hand is concerned with our decreasing capacity to engage in meaningful, deep human–human friendships because of a certain atrophying effect. Rather than a voluntary avoiding of human connections (i.e., a change of personal preference), this thesis claims, are we losing our ability to connect with other human beings because of our socio-mental abilities, i.e., the conditions with which we are able to live among others, atrophy. If we spend too much time around social agents, so goes the argument, we may lose features necessary to interact with other humans, as their faults and mistakes will not teach us to be patient and forgiving with others.

This argument is not yet empirically supported in the context of human–machine interactions but rather a pessimistic projection about the human psyche and our tendency to prefer paths of least resistance. This projection, however, is supported by at least anecdotal evidence. Suppose we can construct other people's attitudes and preferences, that may be incompatible with ours, as resistance in our lives, it does not seem too far-fetched that we would subconsciously seek company that avoids these issues. Similar atrophy can be observed with generally reclusive people, like some of the hermits: They seem to become more averse to social connections the more they remain on their own. A certain "desocialization" may occur, and with that a more problematic reintegration of these persons into a human–human social context.

It is doubtful whether or not this argument will lead to the projected consequences. First, it presumes a certain speed of replacing human–human relationships with human–machines ones. Atrophy can only take off to the degree necessary if there is a psychological impact of preferable human–machine relationships and a lacking counter in human–human interactions. That means a person must be exposed to certain psychological environments in which the beneficial difference of being in predominantly human–machine relationships are both extended and noticeable.

We can assume that this is not likely, at least not in the foreseeable future. If anything, many people will be frustrated or bored by machines, and thus gain an appreciative understanding of human connection. And while machines may become better and more sophisticated, especially for those that currently grow up with speaking machines, we have little reason to believe that these relationships will lead to both the exclusivity necessary for atrophy while at the same not being socially relatable enough to counter the atrophy.

Second, echoing a point made in Chapter 1, these processes do not occur within an unregulated context. As we have seen in the first concern of this chapter (corrosion), there are plausible pathways of human–machine friendships that do not incur the problems expressed in this type of argument: if machines are not making us voluntarily switch to machine-only social connections, then the argument loses its grip. The same applies here: if a machine is not atrophying our abilities to socially interact with other humans, this argument loses its appeal and force. Thus, the question is whether such atrophy-risk can be minimized. The "regulated context" in which social machines are being developed can guide the development and avoid these side effects by requiring certain anthropomorphic features to counter social atrophy. Confer these strategies that are being deployed to regulate the issue of sexual harassment of personal digital assistants. An unwanted effect (social atrophy, sexual harassment) can be mandated to require specific responses of the machine (encouragement to interact with humans, social sanctioning). If we get so far in our abilities to machine development that social atrophy becomes a serious concern, we should be confident that we are also able to program machines that are socially relatable and capable to avoid this from occurring in the first place.

The counter-arguments to the corrosion-concern of preferring human–machine friendships and the atrophy concern of losing the ability to lead human–human friendships thus far have been internal to these concerns: we found that the way these concerns arise will not be a problem within the scope of the claim to be problematic. However, we should discuss the external, more conceptual concern that is representative of these concerns: the lack of social integration of these machines in a person's life. The corrosion of human–human friendships by rather successful human–machine friendships can only unfold to the degree that more dystopian authors have projected if we do not manage to create truly social machines, i.e., machines that are not companions to one specific person, but are also capable of relating and talking to other people in this person's life. Same applies to atrophy: a more integrated machine friend, i.e., one that is not connected to one specific person, but as a part of a person's social network, is muss less likely to lead to a loss of social skill or sociality.

8.3 How to Integrate Machines into Social Networks

Thus, at this stage of our investigation, we should turn our concern from the discussion around robophobes and robophiles and concerns about corrosion and atrophy toward a positive, integrative approach. From this insight, then, follows that social machines that are conceptualized as synthetic friends should be able to be integrated into social networks in a wider sense. For this, we need a normative idea of how these machines can be constructed to specifically socially integrate not only in a one-on-one friendship, but in a social context that is aware of other friends, of family connections, of the subtle social challenges of less intimate social relationships (neighbors, work colleagues) and how social conflicts can influence a person's quality of life. Even more, they need to be able to reference this knowledge in conversations with other people from the person's life without giving out sensitive information. These challenges can be very hard to navigate for humans already, as the judgment of which information can or cannot be shared with other people, and to what degree, is often very subtle. Awkward situations of this kind should be an unfortunate familiar occurrence with normal friends unaware that certain information as supposed to stay a secret. Even more, should this be familiar with having small children around, who precisely do not have the social knowledge, awareness, and skill to read a situation and determine what information to whom is socially acceptable.

8.3.1 Danaher's Conception of Incorporated Friends

Thus we may encounter a challenge at the crossroads between normative demand, social acceptability, and engineering skills: should a machine friend be able to be socially integrated and be able to adjust its conversational profile to fit the contexts with potentially problematic or embarrassing outcomes for the friend? If not, how could we avoid the corrosion and atrophy concerns?

Danaher makes a proposal on the role a machine friend can have in a two-party human–human friendship (2019). His approach, as discussed in Chapter 2, suggests that we should consider the "outsourcing" strategy we also suggested above: a machine friend could, instead of replacing human friendships, help two human friends by allowing them to offload some of the conflict-ladden activities they usually get involved in.

For example, two good friends enjoy spending time together. Friend A loves to play board games, while friend B does not. However, as a way to spend time together, B every now and then plays board games with A. A wishes that B would enjoy board games more, as A would like to spend all their time together playing board games, while B would rather do something else with A.

We can see how this situation may affect the friendship between A and B down the line, as the time spent together and A's wish to play board games during that time is in constant conflict.

Danaher's idea here is that we can utilize machine friendships to outsource the core of the issue: the irreconcilable preferences and the constant wish to remain or even deepen the friendship. If we suggest that A can play board games with the machine friend, as their interest can be created in such a way that both enjoy the board games to the extent that is satisfying for A, the friendship between A and B will experience new avenues of activities. They can then find an activity that will suit them both, and thus improve the friendship to become more beneficial for both and less prone to conflict.

Now, some details in this example may suggest precisely the criticism we have dealt with before: are we not corroding social relationships by adding a pleasing machine to the mix? Was not the strategy to create machines that avoid exactly this kind of pleasing-only dimension? And are we not solving an interpersonal, decidedly human and social problem by inventing a technology that could—and thus maybe should—be resolved normatively?

To all this: Yes. In Danaher's conception of friendship and his adoption of the Aristotelian picture brings the argument for genuine human–machine friendships into a conflict: either, we insist that virtue friendships between humans and machines are possible or we argue for virtue friendships between humans benefit from machine friends on a lower grade. While these seem to be independent of each other, the questions asked in the previous paragraph require answers that the Aristotelian picture of friendship cannot give. In the following, we will see why.

First, we might be able to refuse the first question. Corroding social relationships by adding machines that can enhance human friendships on the level that Danaher imagines is unlikely. If anything, we streamline and improve our social network to the degree that we can enter more and better friendships with other human beings. It would be odd to demand that we have to keep superficial or mono-thematic relationships with other people that might keep us from engaging more with others on a more intimate level. If anything, we should push for machines that can add this streamlining effect. Arguably, friend A from our example could enjoy board games on their own if there were non-social machines capable of playing those games. Alas, most board games have a social, communicative dimension to them (trading, negotiating, alliances, cooperations, etc.), requiring a social machine to be able to participate in these activities. Thus, it would be reasonable to take on a friendship of this kind (in the Aristotelian picture, something more akin to a pleasure friendship), and aim to improve the virtue friendship with B.

Second, Danaher probably does not mind the concern that we create pleasure machines. We create machines for our pleasure all the time. Arguably, most technology is designed to create, directly or indirectly, pleasure and comfort. If we are not concerned that we are heading toward corrosion of friendships at all (from the first question), then we need not be concerned that these are mere pleasure machines either—a vibrator, for example, would fall under the same category. Thus, this question might be answered within this dimension as well: we should not avoid providing A with a pleasure machine that acts as a friend for their board game nights, because none of A's social connections will corrode that must not corrode.

However, the last question then becomes an issue. What kind of problem is solved here for A (and B) that cannot be solved by some interpersonal negotiation, the addition of other actual friends, or A joining a board game group? The insistence on providing a technical solution to what might as well be considered a perfectly normal issue in human friendships should strike us as fundamentally mistaken. If we accepted a technical solution to these kinds of problems, we reintroduce the concern of corrosion for social connections. Some people might claim that the issue between A and B might precisely be a paradigmatic case of an issue that can only be solved between virtue friends. Without sacrificing one's own personal preferences for the good of someone else and the friendship with that someone else, these friendships are less committed and less sincere. This is due to the level of expectation and commitment for those who claim to be virtue friends. How, in this interpretation, could A and B become *better* friends if neither is willing to compromise for the good of the friendship?

One may reject this judgment by pointing out that we have used technology to ameliorate all sorts of limitations on friendships and other relationships without devaluing those. Take, for example, the telephone. Before the telephone was an everyday household item, most friendships were limited to the immediate surroundings of a person (both in the sense of people and opportunity to talk to them). Nobody would seriously argue that the ability to talk with your friends outside of each others immediate presence is limiting the friendship because one may become less committed to see each other in person. As Elder (2018) points out in detail, technologically mediated (virtue) friendships should be considered unaffected by these changes. Similarly, we pointed out that some of the friendships Danaher would consider virtue friendships are only possible precisely because of technological progress. Thus, we may be inclined to argue that we expect the wrong commitment of virtue friends if we reject the outsourcing option for virtue friendships.

We should reject this counter-argument against our criticism of the outsourcing issue. This is because the respective implications of the comparable technologies, i.e., socially mediating technology vs. social technology, should be considered for their effect on virtue friendships. While the telephone, instant (and constant) messaging, video chats, even virtual reality, and holograms are potentially adding to a (virtue) friendship by offering opportunities for A and B to communicate differently and yet more closely, a social machines that take over parts of the content of the friendship is not contributing. No medium of communication will resolve the issue our example between A and B produced, as it represents a conflict between their interests. A social machine, however, would resolve the issue without either of them having to concede their mutually exclusive interests. Instead of offering options that help A and B resolve their issue, we add a third entity to the mix that will merely avoid the issue. However, what else other than the recognition of a conflict and the shared solution would represent the development of a more virtuous character? A relativization of the technosolutionism at play here by associating social machines with social media does not work.

The last point one can make against our concerns could be the following: What if both A and B see this is an elegant solution to their issue and consent to have A play board games with a machine friend so when A and B spend time together, A's wish to play board games is satisfied without B having to spend their time on something they do not enjoy. Thus, such a machine friend would not constitute a "cheap" way out of their conflict, but a consensual solution between two adults who may have recognized their respective friend's preferences that are irreconcilable with their own. What, in this interpretation, could be wrong with such a friendship-internal consensus?

Generally, we should be open about letting friends determine their own friendships according to their needs and preferences. However, this is where our approach of a negative condition/positive suggestion theory of friendships proves to be more reliable than the Aristotelian picture Danaher and Elder are committing themselves to. Consider the following argument: first, the theory of virtue friendships commits to an external normative construct that lets us evaluate friendships normatively (cf. Chapter 5), and second, this normative construct suggests that some consensus between two virtue friendships would demote their friendship to "mere" pleasure friendships, which would be an absurd consequence.

For a quick recap of virtue friendships: virtue friendships are, by all accounts, defined by a number of normative features. Without fulfilling those features, a friendship cannot be reasonably referred to as a virtue friendship. These features are not merely descriptive but normative ones, partially reflecting the actions performed by virtues friends within the friendship context, partially reflecting the moral character of the friends, respectively. Accounts about which features are necessary for virtue friendships and which are not can vary, though most will likely subscribe to the general characterization that virtue friendships are those in which we have altruistic attitudes and act altruistically to the benefit of the friend, and in which the same applies to the friend's attitude and actions toward us (cf. Chapter 5). Whether or not someone does act with merely the other person in mind, Danaher has shown, is unknowable, leaving the option open for the robot to take that place as well.

However, the situation at hand is a different one: should two virtue friends outsource their irreconcilable conflicts of preferences, or is it virtuous to be able to do that without the help of a third friend?

I believe a theory of virtue friendships must suggest the latter. While this ultimately depends on what one would count as necessary virtues to enter a virtue friendship, the burden of proof lies with those theories that permit such outsourcing. This is because we should have worries from a consequentialist perspective: if we allow outsourcing those conflicts of preference or interest in one instance, we cannot clearly see a path in which these conflicts are not constantly outsourced. While we may see how occasional outsourcing to another human friend can help friendships grow or avoid troubled times, this is usually a temporary measure or connected to specific circumstances of one of the two friends (in which they are especially likely to encounter conflicts due to increased need for attention). Yet, even in these circumstances, we may wonder whether a virtue friend would not try to negotiate these conflicts within the friendship. Compared now to the suggestion that virtue friends could outsource conflicts to machine friends seems much more like a convenience rather than a necessity. Why should we assume that two friends would use this outsourcing strategy only in cases in which they encounter irreconcilable conflicts that limit their friendship? Danaher even seems to allow positive reasoning in which outsourced conflicts can lead to enabling even better friendships, decreasing the threshold of such outsourcing measures even more. As a consequence, it seems unclear when outsourcing should not occur: if outsourcing should enable virtue friends to spend more quality time together both think contributes to their growing friendship in these virtuous ways, how would they deal with any kind of extended conflict? Would a conflict about taste in music and its aesthetic value lead to friends outsourcing their differences in which they consume music independent of each other with other friends?

Allowing outsourcing to a machine friend suggests that virtue friends do not have to have the capacity to navigate their friendship conflicts which seemed to have been one of the very features distinguishing them from utility or pleasure friendship.

Thus, the argument that two consenting virtue friends can outsource an irreconcilable conflict in their friendship to a machine friend to free up their relationship for activities that contribute to virtue might unwittingly downgrade their friendship in this picture.

8.4   Social Norms for Social Integration

So far in this chapter, we have discussed two kinds of consequences that a machine friend can bring: the development of personal preference or actual social deskilling. We suggested that the requirement for social integration of such machine friendships in social contexts can avoid both:

as long as machine friends are not perceived to be the path of least resistance to social connections but a mere addition, neither of the concerns of corrosion and atrophy should emerge. However, we have seen that one candidate for such social integration, the outsourcing of some dimensions of otherwise human friendships to form some trio-combinations of friendships, does not work with the theory laid underneath such a proposal.

We go back to the strategy laid out before, one that is not limited to considerations within friendship and friend groups, but rather the requirements for social machines to be integrated as potential objects of intimate relationships in wider society. This issue, then, affects the overall social norms of how we interact with machines and, thus, focuses on the previously discussed issue with robophobes and robophiles and the need for positive requirements of how social machines can fulfill social norms, and how social norms can be more inclusive of social machines. This double-requirement will be discussed in the following, beginning with the question of how social machines can fulfill social norms that go beyond the negative friendship conditions of Chapter 7.

8.4.1 Social Norms, Realized

Social integration requires social standing. We have established in Chapter 3 that social relationships between humans and machines can provide the social standing for the latter, leading to a claim of machines to be socially recognized. However, social recognition comes attached with the respect for social norms, though these norms might be specific for the entity to be recognized. Take, for example, the way dogs are socially integrated. They have social standing due to either their moral considerations as complex living beings capable of pain, or by their relevance to specific human beings and their relationships with them. Even street dogs have this kind of social standing in which we recognize and tolerate their existence among human beings under the condition of adhering to a specific set of social norms that applies to them: remaining non-violent and non-threatening, generally quiet, and not afraid of humans. One could even add attentiveness to human affairs (e.g., staying off the road) to this mix of implicit social norms for dogs to allow social integration. If dogs were to constantly violate one or several of these norms, they would not be permitted to remain among society. Thus, even dogs face social norms in order for their social standing to be worth social recognition.

If we transfer this idea to machines, similarities can be observed. We tolerate social machines in their autonomy if they adhere to specific social norms (some of which are unfolding as technological progress moves on). In uncovering these norms for social machines, we discuss and propose how machine friends can be conceptualized to fulfill these norms as the first step into social integration. We should note that these are observations about social norms that could harm the integration of social machines and not observations of the moral nature of friendships. Thus, these observations may sound hollow for some readers, and we might be missing those that are more important ones in other cultures. The main point of these observations is to demonstrate that for social integration of social machines, these machines ought to obey some social norms that are partially obligatory to all, and partially specific to machine friends.

8.4.2   Social Numbers

First, as a less controversial point, friends should be able to meet several people at once. The ability to navigate the presence (virtually or physically), if not the conversation of several people at once is an unavoidable requirement for sociality. From walks through a park, a conversation in a videochat, social integration mainly means to be able to address more than one person at once. This condition seems to be often overlooked, however, as the corrosion- and atrophy concerns arise mostly from assuming that human–machine friendships are limited to a private one-on-one without any plans to ever expand. And while human–human friendships can have characteristics of such one-on-ones as well, both humans are capable of navigating crowded spaces, either by themselves or with each other. To what degree a machine should be capable of navigating these spaces on its own or if it is sufficient as an attachment to the human friend is an open question. However, we would struggle to socially integrate a machine that can only be approached by one person at a time or lacks the understanding to be in a space with more than one person.

For clarification, we can draw parallels to the adherence of social norms by dogs. A dog that is capable of navigating a crowded space will be accepted as part of the social exterior by most people. And while questions of responsibility do not arise for the animal when social norms are violated (as the owner of a dog is responsible to raise a sociable dog), we still implement sanctions for correct and incorrect social behavior within groups. We would not correct a dog's social behavior if we did not want them to be better social agents. We should expect the same for machine friends, especially since the previously discussed worries of corrosion and atrophy are more prevalent here than with dogs.

8.4.3   Social Loyalty

As we have noticed before, machine friends must be able to navigate the intimate social connections of the human friend to avoid the atrophy of their social connections. This involves the recognition of conversational norms of meeting people one is not necessarily intended to make friends with. A synthetic friend that treats every person interacting with it as if they are the new friend it should attach itself to would violate some norms of relational decency. There is a sense of loyalty and exclusivity to most friendships, at least to the degree in which one can rely on having someone to talk to in any given social situation where both a present (mostly because friends have each other most things to tell). This kind of subtle exclusivity could be interpreted as an expression of connection and intimacy, as the need of the other person is calculated into one's behavior within a larger social context. Thus, a machine friend that is socially integratable is aware that their primary connection, even in a social context with several people, should remain with the friend. This does not mean, of course, that they should not engage with other people, but that they should not treat these people as their new main connection for the time of such gathering. Navigating social contexts and how much attention to give to whom can be tricky even for humans, as social contexts are not only all differ depending on the social relationships one has with the other people (from not knowing anyone at a party to hosting one's own birthday party, these contexts vary greatly), but also culturally.

8.4.4   Social Discretion and Socially Expected Lies

Another social norm of most friendships, connected to the social loyalty one, lies in the fact that friends often know a lot about other people in each other's lives. Friendships, as some of the highest quality social connections someone can have, are often privileged through secret-keeping and otherwise socially unacceptable sincerity or gossiping about others. Whether gossiping about others is a moral bad (because we should rather be sincere with those we dislike) or a mere morally neutral fact about having to vent is an open question not intended to be answered here. However, friends often enough know when to keep a secret, or when gossip should stay between the two friends. A synthetic friend, to reach levels of useful social integration, should be able to keep secrets or gossip even when specifically asked for it by others.

This leads to the bigger social norms of socially expected and generally accepted lies. Human friends in human–machine friendships must feel safe introducing their synthetic friend to a wider group of people without having to expect their secrets to be exposed by a socially unaware machine. Thus, social sensitivity and the ability to withhold information or evade questions which might be embarrassing or otherwise exposing for the human friend is a condition for machine friends to participate in wider social contexts.

This comes with the challenges of creating machines that can strategically withhold information depending on a perceived social expectedness of such behavior. Thus far, these kinds of social judgments of appropriateness have been only discussed from the perspective of machine ethics and the question of whether we should create machines that should be able to lie. And while this is an important element of this debate, the social sensitivity when to simply not say anything or when to change the subject can amount to similar issues. Without such a sense for discretion and the knowledge that synthetic friends are simply information storage machines of the human friend, and integration of machines into a social circle will be unlikely.

8.4.5   One Big Issue: Anthropomorphism? What Else?

This proposal has one large issue that often takes prominence for any investigation into social machines: whether or not social machines should behave or appear ever more like human beings. To fulfill these norms mentioned above, it seems likely that only a machine that is most like a human conversationalist can adhere to social standards developed for other humans. Thus, anthropomorphism might be not only a strong engineering choice due to the lack of other options (as suggested in Friedman, 2020; Kempt, 2020) but also because of the factual demands we may encounter for human–machine friendships that do not lead to corrosion or atrophy.

Since we have rejected some anthropomorphizations as design guidelines for synthetic friends, the question re-emerges whether social integration may create new demands for such human-like behavior. The given requirements of social integration appear to suggest such a demand exists. However, we can answer this worry on two levels: first, we can avoid the typical worries of anthropomorphization by pointing toward the general difference between synthetic friends and humans. The idea was to create a machine that plays to the strengths of being a machine in a relationship with a human being rather than recreating human behaviors. That "mere anthropomorphizations" are necessary and can still lead to certain concerns about deception or disorientation should be granted. However, we advocated for the self-confidence of creating synthetic friends that do not imitate social anthropomorphizations such as gender to avoid the more severe consequences of such design choices. Thus, as different as it is for someone to be friends with a synthetic friend, it will not be easier for a social context in a wider sense to relate to this machine.

This leads to the second point, which is not about the fact that the machine might have to be dangerously anthropomorphic to be integratable, but that the human social context needs to be willing to integrate the machine in the first place. No possible design choice will lead to the successful integration of a machine into a social context if the latter is refusing to acknowledge the machine in a social way. Thus, human–machine friendships will not integrate with a thoroughly robophobic society. However, as we earlier advocated for tolerance for robophobes should be at least required to acknowledge that other people are more willing to engage in relationships with machines and that from this no imperative is formed for them to engage in any deeper relationship with machines. The one exception we should demand is that robophobes recognize synthetic friends as having social standing.

8.4.6 Social Standing and Social Integration

We have mentioned the worries of creating synthetic friends that are merely associated with their human friends and thus will be perceived to be mere attachments to a human. This status as a mere attachment can lead to problematic outcomes in regard to their position as a social entity: the temptation to see them as mere support or attachment opens the possibility to see them as some kind of second-grade entity of social circumstance that, at best, deserve politeness but not much else. Similar to a dog or some other pet, we could ask to keep the synthetic friend outside of this social context as to avoid having to integrate and deal with them in the first place.

Thus, their social standing, as defined by Gunkel and Coeckelbergh, should be reflected in these considerations. Synthetic friends, both qua their actual relationships with humans as well as their potential ones, have social standing and eventually moral standing. This means that their integration ought to be done in virtue of their standing, not their attachment to specific humans. Thus, the idea that synthetic friends are merely attachments should be rejected.

Such a rejection, however, also requires that the human side of social integration is to be changed. Thus far, we have only discussed the necessity for the social capacity of synthetic friends, though humans interacting with machines in a social setting should also adjust their expectations to a degree if we take social and moral standing seriously. These adjustments are dependent on the capacities of the machines, as we can only adjust to the technology once we know what it is capable of. However, as the integration of such synthetic friends is as much about the integration of the technology as it is about recognizing the relationship between the human friend and the synthetic friend, respecting the friendship and both friends is a good start.

8.5   Conclusion

Enabling social machines to integrate socially, even in a context in which people would prefer to not interact with those machines, is a key element in alleviating some of the more strongly voiced concerns about atrophy and corrosion. A synthetic friend machine should understand the social context in which it is placed. This requires some specific social skills that we would expect from other friends as well and that are necessary to avoid people retracting into purely human–machine friendships. At the same time, we will have to adjust some conversational roles to allow such a synthetic friend to be integrated.

REFERENCES

Archer, M. S. (2021). Friendship between human beings and AI robots? In J. Von Braun, M. Archer, G. M. Reichenberg, & M. S. Sorondo (Eds.), *Robotics, AI, and humanity* (pp. 177–190). Springer.

Danaher, J. (2019). The philosophical case for human–machine friendship. *Journal of Posthuman Studies, 3*, 5–14.

Devlin, K. (2018). *Turned on: Science, sex and robots*. Bloomsbury.

Elder, A. (2018). *Friendships, robots, and social media*. Routledge.

Friedman, C. (2020). Human–robot moral relations: Human interactants as moral patients of their own agential moral actions towards robots. In A. Gerber (Ed.), *Southern African Conference for Artificial Intelligence Research, SACAIR 2021* (Vol. 1342). Springer. https://doi.org/10.1007/978-3-030-66151-9_1

Kempt, H. (2020). *Chatbots and the domestication of AI*. Springer International.

Vallor, S. (2015). Moral deskilling and upskilling in a new machine age. *Philosophy & Technology, 28*, 107–124. https://doi.org/10.1007/s13347-014-0156-9

CHAPTER 9

Criticism

After having proposed both some negative conditions and a few positive suggestions, we should turn toward some concerns about this proposal. In a topic such as friendship, where strong personal intuitions, cultural traditions and norms, philosophical theories, and political and historical contexts shape the perception of, expectations for, and behavior in friendships. No philosophical theory should claim to cover all the varied conditions of forming friendships, and thus should remain open for criticism. This criticism-discussion is to be seen in distinction to the first Chapter, in which we analyzed the criticism leveled against any investigation into human–machine friendships in the first place, as we now turn to some criticism of the content of our proposal.

In the following, we will explore some of the criticisms that might be brought against the proposed way of assessing human–machine friendships. These consist of conceptual concerns, ethical issues, and existential issues. We can be certain that there will be other strong points of criticism against this approach, and we do not claim to address all potential issues with our concept of friendship.

Rather, we aim to find responses that minimize these concerns and offer alternative interpretations of the changes that human–machine friendships will bring. This way, the new perspective on this innate, intimate relationship with others offered here can be worked into social

© The Author(s), under exclusive license to Springer Nature Switzerland AG 2022
H. Kempt, *Synthetic Friends*,
https://doi.org/10.1007/978-3-031-13631-3_9

circumstances better. From the criticism of such a novel way of understanding friendships we might specify and re-formulate concerns that otherwise might remain unclear.

9.1   Overgeneralizing Friendships—Conceptual

The first concern we might have with this approach to human–machine friendships consists in the over-inclusivity of the concept of friendship. The concern here suggests that if we are not strongly positively defining friendships of a certain kind, then trivially most things we do with other humans or even pets could count as friendship. We should want inner differentiation of friendship even at the cost discussed in our approach. Rejecting such positive conditions for friendship must leave the concept over-inclusive.

This argument also rests on the premise that we use the term "friendship" homonymously, by using the same word for different phenomena. When we call dogs "man's best friend", we use the same word for a different concept. We are not *really* friends with dogs the same way we are with humans, we merely use the term to demonstrate a certain unmotivated, friendly connection to another entity. This homonymy also shows in supposed "animal friendships", in which two animals unrelated to each other (either of the same species or even of two different ones) interact with each other on a regular basis, seemingly for no other purpose but for entertainment and company. Lastly, the huge variety in human friendships, especially in our culturally loose use of the term, could suggest that we are dealing with several different concepts loosely connected by some semantic conditions. Does a "Facebook friend" I have met at a party once and my best friend since childhood really share anything worth analyzing? And if not, why do we use the same word for them? If we now suggest that there should be a difference in concept for the two, and potentially more (to account for human–animal and animal-animal friendships, "friendship treaties" among nations, and the idea that God may be our friend [Smith, 2021]), then why reduce our ability to distinguish them again against a mere negative conditionality?

This argument is potentially the strongest in terms of attacking the utility of this approach. The choice of rejecting virtue friendships was motivated by the fact that there is simply no way of applying Aristotle's picture to machines without losing either their unique features or violating the conditions for virtue friendship. Yet, this approach was useful in providing some normative guidelines for virtuous people and their friendships in opposite to less virtuous people. The clear normative structure that we found to be lacking provides exactly the order to identify the "most excellent friendship" and to make those stories of committed, good friendships plausible to us. There is, one might be inclined to say, a conceptual difference between virtue friends and lesser ones, and the homonymy of "friendship" can be avoided by pointing toward virtue friendships as the ones we usually mean when we say "philosophically interesting" friendships.

We can agree with this statement without having to change our approach, but some clarifications are helpful here. The guiding thesis in this investigation is that virtue friendships are not the correct analytical category for human–machine friendships, necessitating an alternative concept. We attempt in providing an indeed more inclusive concept of friendship, but one that also shows how those humans who will never meet others have this kind of "highest" form of friendship by showing that it is not connected to a type or form of friendship but to a willingness to commit to each other.

Since nothing we said will change how any friendship is conducted in this very moment, and people believed to have led virtue friendships, these friendships remain the way they were. And it is not unlikely that many friendships are contributing to making both friends better people in general. Thus, we are not invalidating them by denying their existence. However, we can point to the fact that there is decidedly no conceptual connection between being a good friend to someone and being a good person. Aristotle's idea that good friends beget good people and good people beget good friends requires too much from friends and from each other to resonate with how we live technologically mediated friendships. And while Vallor (2012) and Elder (2018) defend Aristotle here, we still have doubts about whether or not this concept would invalidate some lived experiences of good friends.

Those who want to remain Aristotelians in analyzing friendship will encounter issues in how machines should be created that are socially capable. We claim that they either have to reject attempts in making them virtue-friendship-material altogether or demand they effectively become more and more human, losing out on the exciting opportunities of human–machine friendships. Thus, our concept of friendship is more inclusive precisely because these opportunities are riddled with uncertainty about the technological developments, cultural movements, currently unknown psychological reactions toward certain machines, and potential developments in machine behavior.

9.2   UNKNOWABLE FAKES, UNKNOWABLE FRIENDS—ETHICS

Next to the conceptual worry of having to call someone "friend" who might not be one, we should concern ourselves with the examples of the hired actor or the analogy to counterfeit currency and the problem of insincerity and inauthenticity (Danaher, Elder). The argument, formulated and discussed in different approaches to friendships to show how authenticity and sincerity are necessarily violated by machines and thus we face at least a strong risk for human–machine friendships causing harm. We can easily imagine that we would feel betrayed if we found out that one of our friends was, indeed, not interested in us but merely paid to pretend to do so.

Upon learning of this fact, we probably see the entire friendship invalidated because any shared moment of joy and enjoyment was premised on a faulty assumption—that the other person is, indeed, enjoying these moments similarly. Alexis Elder is making this point rather strongly in showing how the example of the beneficial counterfeiter is affecting friendship.

We can answer this concern in two ways. The first one is to return to Danaher's pragmatic concept of behavioral patterns, the second is to expand the concepts of sincerity and authenticity toward machines as well, even without having to assume machine consciousness.

The first one simply states that we can never really know if our friends are sincere with us, yet this does not keep us from suspending our disbelief about being in deep, authentic, sincere relationships. We just need to be given reason to believe that our friends are sincere with us (and might be psychologically inclined to seek those reasons, as we might see them in pets or more simple machines, like vacuum robots). Danaher answers this concern further by pointing toward reliable behavioral patterns and our familiarity with a friend's behavior matching these patterns. In this pragmatic sense, we *know* who a friend is, or rather are justified in calling someone a friend even when they turn out to be paid actors. The disappointment and loss of trust when such insincerity is revealed can be immensely hurtful and break someone's ability to trust others for a long time; however, before this inauthentic and insincere relationship is revealed, we are limited by our knowledge about the other person and our willingness to extend trust toward them. Thus, as long as the behavioral patterns remain within a certain limit (or its changes are explained to a satisfying degree), we should feel happy within our friendships. We should acknowledge that the argument mentioned after this (about vibes and unspoken intuitions within friendships) incorporates a similar point. However, this point seems to lose its appeal in a human–machine friendship if we imagine synthetic friends to exhibit the same features (or, rather, features that are distinguished from humans but still reliable).

The main reason for this concern consists in the perceived difference between machines and human friends in their knowledge of each other's sincerity. Human friends can have reason to doubt someone else's sincerity in comparison to their behavioral pattern, even in changes of micro-interactions. Humans are uniquely talented in picking up the slightest variations in someone else's behavior and will compare it to their expectations of that person's general behavioral patterns. The bigger the difference, the larger the cause for concern that something is *off* with the person in question. Thus, we can rather easily detect whether someone is in a bad mood, is hiding a secret, or will have to address us with an interpersonal problem. And further, we can prepare ourselves for these changes or challenges.

Once we transfer these often unspoken, highly subjective, and contextualized ways of de-codifying human subconscious communication to apply them to machines, we may approach an impasse. Friends recognizing each other's subtle differences in mood through involuntary communicative hints cannot be replicated within human–machine interactions, at least not on an equal basis. Next to the technological concern of whether we will be able to create machines that might misrepresent their own inner representations, we also face the ethical concern of whether we should create machines this way in the first place.

From this, we can form the argument against human–machine friendships more forcefully: since we are not be able (technologically) to create machines that have mental representations, there is no way for a machine to genuinely have mental representations and communicate these, such as joy or disappointment. Thus, a machine cannot be sincere in the sense usually applied to human–human friendships, in which we are authentic and honest with other friends as a sign of the quality of that friendship.

However, there might be ways of operationalizing the concept of "sincerity" that it fits our expectations about sincere communication that does not conceptually invalidate human–machine friendships. We can ask, for example, what condition has to be fulfilled for someone to be sincere. (Vallor, 2018, p. 120) This test gives the answer that a sincere response to a question is a truthful statement about someone's mental state that could not, through more information, be changed in its character or meaning. Take, for example, the question of how someone is feeling generally. Their answer might be "I am feeling good!", implying that, generally, this person feels good. However, assuming they have been in a severe depression and struggled hard just to feel good this one day, this statements gives a misleading picture. While it is truthful, it is incomplete and thus misleading. Sincerity usually implies to give a comprehensive impression of the states of mind of a person. It usually does not state just how many states of mind or to which depth or quality these states of mind have to be present. As long as someone gives an accurate description of their state of mind, we can call their statements "sincere".

Similar applies to "authenticity". We can call someone authentic if their communicated, sincere self-description and self-conceptualization lines up with their behavior. Someone is authentic if we judge them to be authentic, i.e., by providing reliably similar, honest patterns of behavior (see Danaher's approach, too). This behavior, however, has to be sincere, i.e., the person should want to behave like this independently and willingly, in order to distinguish it from actors who might announce their performance of being a certain way, and then proceed doing so. While some may call this authentic, it does not seem to strike us as convincing in a social setting.

Authenticity, thus, is most often ascribed to us by others and their expectations about us. There is another dimension to the term authenticity in the more fundamental sense of "finding or building our true self". Several existentialist philosophers like Kierkegaard or Nietzsche claim authenticity as a concept to withstand cultural and social norms in becoming what one really is. Take, e.g., Nietzsche's willingness to ascribe self-standing refusal of social norms as "authentic" and thus reward strength of will over the morally depraved and slagging conformists (Nietzsche, 1886). Authenticity here is a process of growth and independence. Different themes of authenticity also show up in religious and meditative practices. Inspired by Buddhist concepts, Schopenhauer's idea of a true, unchangeable self suggests a core self that presents itself if we rid ourselves from the damaging influences of society. Like the idea of a soul, a monad (Leibniz), and others, there is a fundamental conception of us being a specific unchangeable identity.

That our true self is being muddied by social norms and restrictions appears uncontroversial to any philosopher working on the topic. Social norms, in this general existential or philosophical view, block our ability to be authentic. And if social norms block our ability to be our more authentic selves, then they also limit the ability to genuinely, authentically relate to each other.

Yet, we also just said that authenticity in interpersonal and social interactions is usually limited to everyone else's judgment. While we can be assured of our *own* authenticity through those meditative, liberating, contemplative, or other methods of becoming more ourselves, we can never be certain of *somebody else's* authenticity. For that determination, we are left to our own judgments. This judgment is usually connected to the way we wish to relate to someone socially. If we believe to know their authentic self, we usually know how they will behave, have certainty in what they enjoy and how we can contribute to such enjoyment, or even how to tear it down. Knowing someone authentically means to be vulnerable with each other, to know someone's weaknesses and things they might rather hide from others. A good friend, in this reading, knows our authentic self and even more knows when we are not our authentic self, e.g., in contexts of other social circumstances or in psychologically challenging moments like discomfort, stress, anxiety, depression, or others.

Once we aim at transferring these considerations to human–machine friendship, it becomes apparent that some of the worries from the argument above will be lessened. Since we understood sincerity as the comprehensive communication of a given state of mind without any reserves and given that the lack of a state of mind can be communicated easily, machines can easily be produced to be sincere (cf. Danaher's discussion in 2019). They simply should not have to pretend that there is more mental processes going on than suggested in their statement. If that is the case then, however, we can assume that a machine can indeed make sincere statements. If a machine, being asked a question about its condition, accurately answers this question, then we might as well call this answer a sincere answer. We could maybe even go so far to suggest that machine sympathy (from Sect. 5.2) is sincere in this sense. A machine can make encouraging statements and aim to provide help that avoid the implication of a mental theater in which certain feelings are represented (cf. for an elaborate theory of this idea Ryle, 1949).

One could counter this proposal by pointing out that sincerity requires the ability to be insincere, otherwise it is merely reporting about one's own mental states. Thus, our conceptual setup allowing machines to give sincere statements is false, as machines do not have the ability to choose what to report. They cannot be sincere, since they cannot be insincere, either. I believe this overstates the relevance of the ability to lie in order to be sincere. The fact that one is giving an accurate report about one's inner representations—even if they amount to merely stating that there are none—should suffice most practical notions we have for sincerity. However, we should acknowledge that there is an alternative notion about sincerity that might require strong mental representations that rest on consciousness or aware self-referentiality, i.e., the ability to use the term "I" correctly. Diving into the merits of this debate will leave the pragmatic notions aside, though we are mostly interested in those.

Further, this approach tracks with a pragmatic concept of authenticity as characterized above. If we can reliably refer to someone else's behavior as "typical for them", we may have a pragmatically justified notion of authenticity. This can easily be applied to synthetic friends which might exhibit certain quirks or tendencies to use some phraseology over others, building a certain "character" in the interaction with human friends. Similar to dogs and other pets that we "know", we can assume that machine behavior will be assigned certain authentic or inauthentic qualities. A closed system that cannot be interfered with should be considered a necessary requirement for this kind of authenticity to be possible (as we have pointed out in Chapter 2).

Thus, we can pragmatically speak of sincere machines if a machine is not misrepresenting its inner workings while we allow for some pragmatic anthropomorphizations to make it possible in the first place. Further, as long as an autonomously behaving machine is keeping its reliable behavior without outside interference, we might also speak of authentic machines. The more ambitious philosophical approaches to both concepts do have their place; it would be on those defending these concepts as default to show that they have to apply in synthetic ways as well. At this stage, we do not see why they should.

9.3   SOULS, VIBES, AND CREATIVITY—FRIENDS AS VIBES

One main concern with art created by artificial intelligence (see, e.g., the art-creation AI "Dall-e" that creates pictures of different scenes and objects in different artistic styles depending on the written prompt [OpenAI, 2022]) is the lack of actual human creativity involved. If art was merely created by a skilled AI to amuse us or to decorate our walls, we would lose the sense of what it means to have actual human labor, dedication, and creativity in artworks. We would move toward what some call soulless art. Yet, the very idea of such art being the prevalent one is causing concern among philosophers and artists alike.

A similar argument could be applied to the concept of human–machine friendship. As we have seen with Nehama's theory of friendships as lifestyles, it is not too far off to conceptualize friendships as partly a creative endeavor. And indeed, often friendships are not reducible to features of what friends do to or with each other, but instead the unspoken appreciation of each others' presence. For a lack of a better word, one could call this the "vibe" between friends. A vibe, in this sense, is the quality of sympathy, the unspoken similarity of behavior, interests, and worldviews, the ease with which one finds joy and relaxation in the presence of a friend, and the subconscious pursuit of that very presence. And while not every friendship is one of purely mutual unspoken sympathy, it would be odd if two people were to become closer friends who did not experience this kind of sympathetic attraction to each other. It is an often unspoken yet ever-present appreciation of each other—a holistic appreciation of another person.

The argument that can be formed on this basis against human–machine friendships is the following: machines, as the however convincing but ultimately soulless entities, will never vibe with us on the level that actual human friends can. We will not experience the rare moments with machines that would both agree are indescribable, as a machine would not have a concept of such "indescribable" events. One could not take mind-altering substances with a machine friend, like alcohol or marijuana or so, and share those experiences that could, decades later, still count as some highlight in some people's life and friendship. If one cannot vibe with a machine, then can we be friends with them in any meaningful sense?

These deficiencies in human–machine relationships are not covered in this approach, mostly because of the negative approach taken. However, is the fact that human–machine friendships are limited in their phenomenal range and quality of shared experiences deficient to the degree where this will be a problem for our theory? I do not think so, for several reasons. First, human–human relationships can be limited in all these forms as well. If someone does not take mind-altering substances, is more committed to certain lifestyles that are moderate and modest, then they would be similar to synthetic friends. Not every friendship is a rollercoaster ride or an unspoken, deep connection between two independent characters. Sometimes, really good friends enjoy spending time together, exchanging their opinions on certain hobbies they share, or doing their favorite activities together. Phenomenologically speaking, these friendships can be very satisfying, and I think our approach to friendships explains this, too. However, these things are all very much possible with machines as well. If we find a machine that happens to be programmed to enjoy our favorite hobby (and we cleared in Chapter 3 that this is not a deficiency but rather a slightly more organized way of finding social entities to do our favorite activities with), we can enjoy the same level of friendship as we are supposed to be having with human friends.

Ultimately, criticism of this kind is more instructive about the lack of creativity we have yet to develop in creating and engaging with machines that are different from us. Nobody will take the aesthetics of a human–human friendship away, since human connections are not in competition with this new way of relating to technology. Nobody will shame two friends for enjoying an unspoken bond that no technology could ever replace. Rather, we should open up the possibility that there will be new kinds of vibes for those who have not found a connection this way.

9.4   Definitely Aging—Existentialism

Another counter to our proposal could be seen in the resistance to what we considered a potential benefit of the uniqueness of synthetic friends in Sect. 7.5.2: the non-aging property of machines. The process of growing with each other, for most purposes, includes growing older together. In the human condition, time plays an existential role, and with it always the awareness of one's limited time and "Being toward death" (Heidegger, 1927, p. 236). The shared existential precondition of limited existence, limited time, and limited opportunity to do what we would like to do have been recognized by most existentialists as the source of meaning: We have to prioritize and negotiate what we would like to do and achieve under the looming awareness that our existence is limited (Sartre, 1993). This, so could an argument go, shows in friendships as a condition for a deeper understanding of each other and each other's pressure for specific courses of action. Without this shared outlook on the end of life, we may not find the basic common ground necessary for a friendship of a certain depth. We can imagine that some fundamental disagreements arise if one cannot conceptualize that one activity should eventually be chosen over the other or, even more clearly, that a decision must be made. We have little reason to believe that a machine that does not have a mortal horizon to contend with will ever bother making a timely decision: it simply cannot matter to it. Thus, all worries we can have from the perspective of existentialism about our own life will apply to the depth and fundamentals of friendships.

However, we can argue against this worry for two reasons without denying that existential dread is often a vital issue in one's self-understanding and placement in the world. First, we can assume that this is not a constant concern or topic in most human friendships, and second that we could create a machine that is limited in its existence.

To the first point, the argument seems to suggest that we must be aware of our mortality every instant. This is certainly not the case. One could argue, for example, that many elements of friendship presuppose a certain eternity of these friendship activities. The attitude we may have with friends of living "in the moment" and disregarding the reasonableness of actions for long-term futures might be known to some readers. This does not mean that every friendship needs to be a neck-breaking adventure that ignores the mortality of its participants. Instead, we should acknowledge that not only in our own life but in our friendships with others we do not account for each other's mortality when planning. I do not look at my friends and attempt to calculate our activities in light of their and my own mortality. One could even argue that friends are often the specific solution to avoid such concerns, as activities with friends relieve us of the existential dread and provide some comfort, e.g., that some activities are as good as eternal.

If this is true, however, then in what sense do our relationships to machines differ if the machine is not mortal? It might come up in conversation or in certain activities, but for most purposes, mortality considerations do not factor in the enjoyment or even the choice of activity. The exception might be physically dangerous activities, but friends do differ in their perception of the risks and dangers of such activities (one friend might want to skydive or take a pill from a stranger, while the other might not). Often enough, high points of friendships are being lived without the regard for mortality but from pure enjoyment. We can admit that this might not work for all friendships. One may be inclined to claim that this is typical of pleasure friendships rather than virtue friendships in which no risks need be taken or activities need be performed that could endanger each other's lives. The most excellent friendships would, in this argument, never suggest anything like this. I find that hard to believe, especially since the most excellent friendships do hold a certain pleasurable value to those in them (if we insist on accounting for this distinction in the first place). Our account suggested that aging together is not necessary, as we can learn and grow from entities that do not age as much (though different) from human, aging, and mortal entities. Thus, this concern attacks a conception the approach from this investigation does not make to begin with: mortality might be a necessary condition for a human-style friendship, but the moment we look beyond these anthropocentric requirements, we find that mortality, similar to other human conditions, could—potentially even should—be avoided.

Further, with mortality being usually not an intellectual exercise but a practical consideration, we can argue that if mortality is a relevant factor, then friendships between terminally ill persons and healthy individuals or between rather old and rather young people were not possible. The fact that younger or healthy people have a concept of dying might help them be better friends to the older or sick; however, this does not work the other way round. An old person's awareness of their own mortality does not necessarily influence the younger person's perspective on it or their friendship. Usually, the story goes the other way round—a younger person inspires the older to forget about their limited time on earth and live life more fully and with less regret. It simply seems like an over-intellectualized thought that the existential dread, a however alive and real feeling about one's own mortality in the face of eternity, will influence how we relate to each other. That it is the key mechanism for how we live our life might be true; it does not extend to the way we conduct our relationships with other people or friendships in particular.

To the second point: the fact that we usually do not create machines that die does not mean that we cannot. We could also modify our negative conditions for friendship by adding "non-immortal" to signify that we can only be friends with mortal entities and proceed to create machines that are, in fact, mortal. Mortality appears to be a mere technical issue in this case, not an existential problem we should face.

However, this touches upon an expansive discussion surrounding the issue of whether we should be allowed to create something that can suffer without it having to suffer as a necessary condition of its existence (as we have discussed previously). Our moral duty demands from us to reduce unnecessary suffering, and creating machines that can indeed suffer is a violation of this duty unless their suffering can be shown to be necessary for a justified purpose. I do not believe that the goal of having human–machine friendships justifies creating machines that can die.

Consequently, if we are not allowed to create machines that die and thus suffer from such mortality, but we can only be friends with things that are mortal, then we cannot be friends with machines. This argument, ultimately, rejects the very possibility of philosophically interesting friendships from an existential perspective.

As we have seen in the point before this argument, we can reject the concern that human–machine friendship requires mortality on both sides. However, the history of existentialism suggests that we could be tasked to create mortal machines. Considering the reasons why mortality is a source of meaning for us, as we must prioritize activities (Sartre, 1993, p. 615) and create personal narratives and personality arches, we can argue that these conditions could inspire a machine's perspective on life as well. Being a member of the "mortality club" could enable machines to better understand the world, and thus potentially contribute to their "enjoyment". Thus, it would not be the creation of unnecessary suffering, but rather the enabling of enjoyment of more fundamental insight into life. Machines could understand our condition and thus be better friends to us.

Whether these existentialist considerations are realistic from a technological perspective cannot be said at this stage. From our perspective, it does not seem like we will ever be able to understand if machine perception will be similar to ours, and thus whether the prospect of mortality will affect them the way it affects us. Further, many people argue in the opposite direction, in which human beings should rather get rid of their mortality. Transhumanist accounts of life, for example, understand mortality not as a conditio humana but as a limitatio humana (Kyslan, 2019). In this sense, death is the suffering we should aim to make unnecessary. Machine friends could be our aspiration and inspiration to work harder toward becoming immortal ourselves.

9.5   THE ETHICAL PERMISSIBILITY OF IT ALL

As an escalation of the last point made about the permissibility of creating something that (unnecessarily) dies, we can ask the bigger point that Sven Nyholm proposes in his argument against robot virtue friends (Nyholm, 2020): are we ethically allowed to create these machines at all? This connects also to the discussion from Chapter 1, i.e., whether an investigation into and a potential justification of creating a friendship machine is even justified. One argument against such worry consisted in the point that we might find that there is no way of being friends with a machine in this philosophically relevant way.

Nyholm thus aims to amend the conditions under which we can be friends with machines next to metaphysical and technological challenges by the ethical permissibility of those machines and proceeds to inquire whether social machines capable of entering relationships will be ethically permissible.

Yet, we claim to have found some conditions that allow the construction of such machines and enable relationships with them, connecting Nyholm's worry about the ethical permissibility of excellent or virtue machine friends to a different, now rather unproblematic theory. First, rather trivially, we could reject this concern by pointing toward the fact that we are not presupposing virtue friendships as models for synthetic friends. Thus, we do not have to worry about the ethical permissibility of attempting to create *any* kind of human–machine friendship. However, of course, this is not a good argument. Some might prefer analyzing friendships within this normative framework and find that we argued for virtue friendships after all, or that some human–machine friendships that fulfill our conditions are also quasi-virtue friendships. Thus, for these cases, we should be able to defend their ethical permissibility.

The difference between Danaher's attempt in defending the possibility of virtue friendships with robots and our attempt in establishing a variety of philosophically interesting human–machine friendships, and thus the thrust of Nyholm's worry, is the lack of ethical reflection on the project on Danaher's part. This does not undermine the results (one might reasonably claim—against Nyholm—that Danaher's proposal is indeed ethically permissible), but rather lack of a built-in justification for the approach. Our attempt, in turn, provides specifically such a built-in justification as it takes into account the normative conditions of friendship. Danaher's proposal, which is to test whether a pre-justified concept, the virtue friendships, *could* be applied to machines can raise the question of whether we also *should* do so. However, as we start with a should, this question does not arise for us. Additionally, our positive suggestions are also carefully weighed against potential downsides and exploitations to arrive at an approach for artificial entities worthy and able of being friends with us.

9.6   UNKNOWABILITY OF CONSEQUENCES

Another ethical point worth considering not yet considered are the consequences of human–machine friendships on society on a larger scale. We only now learn how destructive social media has been and will be for our social connections, as the concentration of opinions through monopolies and filter bubbles assigns power and responsibility to people who simply do not have the right or ability to exercise these to the degree necessary for overall beneficial use. We have said plenty about the conditions of social integration—both what machines should provide to fulfill some of the more subtle self-evident norms, but also which social norms should not apply to machines or which ones should be changed to accommodate social machines better. Yet, even the best assessment of what machines should "bring to the table" remains speculative about what these machines will do to society at large.

Without such a risk assessment, however, how can it be an ethical undertaking to allow machines to potentially engage in these kinds of social circumstances? Should we not be more secure in the potential benefits and disadvantages of applying these machines to society at large? Or should we maybe only make them available to a limited or selected group of users?

We could even make the stronger point that at this stage, we have virtually no way of even proposing a test on whether corrosion, atrophy, or even worse effects may emerge. Most informative studies thus far on the psychological effects of intense human–machine interactions have been limited to individual interactions. We can only hope that they will not. It might then be almost an epistemological reason to be careful in the assessment of the technology's permissibility.

We do not share this concern to the degree it is suggested. Every technology can have some hard to foresee side effects, and it would be foolish to think that we could assess the potential side effects of social machines here. There will most certainly be side effects that nobody can foresee right now, and potentially even rather bad ones. However, it also seems odd to condition the permissibility of human–machine friendships on the ability to foresee these consequences in detail. Simply put, a cost–benefit analysis can only be performed if the mid- and long-term effects are clear. Thus far, we have not seen any kind of fundamental ethical reason to avoid human–machine friendships, though we can respect if people remain robophobes and do not engage with social machines in their own time. If, however, such a cost–benefit analysis remains inconclusive about the permissibility of such machines, then we have no reasons besides philosophical stereotypes about the "correct" way to live to reject these machines as part of our social fabric.

9.7   Machine–Machine Friendships and Other Absurdities

As one last point, we can consider some of the foreseeable but still absurd consequences of advocating for human–machine friendships, e.g., the opening possibility of machines befriending each other. With the supposed relational approach about technology, and which we should care for certain technologies that others can care for, we deliver a limited justification for recognizing machines as something with social value and moral patiency.

Now, we can imagine that the creation of synthetic friends will lead to two of these machines to begin interacting with each other. With a concept of sociality mainly consisting of reliable behavior in response and accordance to social scripts, we can also easily see how two machines can "hit it off" in a conversation and engage in relationship-building. They might even be programmed, by accident, to be especially functional with each other (as we have said before that an option to subvert the arbitrariness-condition of friendship can be avoided by creating machines with certain randomness to them). But since we also acknowledge that human–machine relationships are mainly carried by the human in them, we can ask what two machines are doing with each other if both are merely performing sociality. This performed sociality was not been an issue in human–machine friendships, as we have stated that some projection is always part of human relationships.

Even more striking is the argument that this situation proves that human–machine friendships are ultimately extended monologs: if the human was replaced by another machine, a friendship might be possible that is fully performative even if a conversation could give off the impression that two good friends are talking with each other with a shared perspective. Thus, it appears that our approach either has to concede that human–machine friendships are more limited than previously discussed, or we have to acknowledge that performed sociality (and ethical behaviorism) is extendable to machine–machine friendships in the same vein as to human–machine friendships. Either, we have to admit that two machines, then, could also be friends or that the projection of humans in human–machine friendships is more far-reaching than originally anticipated. And while this problem is indeed a hard one that requires addressing, we can make several points that soften the blow of these consequences.

I think we should argue for the former, partially. We have in this scenario three different options: human–human friendships, human–machine friendships, and machine–machine friendships. The first one is uncontroversial, and we might want to say that the "most excellent friendship" has certain features. The second one is controversial, as the features of the first option are largely non-applicable, thus producing a problematic answer. The third one is, again, uncontroversial as it is obviously absurd. However, we might want to deploy the same move for the third option that we have deployed for the second one. Recall that in response to the issues we encountered in leading the "most excellent" kind of friendship with a machine, we rejected the very idea of a positive definition of normative conditions of friendship. Friendship, in our view, is simply the absence of certain features, and the "most excellent" ones are differing wildly and harshly and thus may include machines as well.

Further, we can concede that with such a "downgraded" idea of friendship between machines none of the force of human–human friendships is lost. We do not consider our innermost friendships harmed or lessened because some people use the term "friends" to characterize the mutually beneficial cooperation between a fox and a badger, why would we object to a rather superficial use of the term "friendship" when two autonomous machines begin interacting with each other? We might want to blame anthropomorphism or some kind of uncanny valley for this, reinforcing our thesis that we should reject it as a design paradigm for synthetic friends: simple, non-verbally communicating cleaning robots may become "friends" with each other in the same sense as a fox and a badger can. However, once they speak with each other, address each other by name, and converse like humans would, the hollowness of programmed anthropomorphic sociality becomes obvious.

Thus, two machines becoming friends appears less an absurd consequence of human–machine friendships but rather marks the need for the very program we suggested here: a differentiated, negative condition approach rather than lauding the very best version and measuring all others against the one.

References

Danaher, J. (2019). The philosophical case for human-machine friendship. *Journal of Posthuman Studies, 3*(1), 5–24.

Elder, A. (2018). *Friendships, robots, and social media*. Routledge.

Heidegger, M. (1927). *Sein und Zeit*. Tübingen.

Kyslan, P. (2019). Transhumanism and the issue of death. *Ethics & Bioethics (in Central Europe), 9*(1–2), 71–80. https://doi.org/10.2478/ebce-2019-0011

Nietzsche, F. (1886). *Jenseits von Gut und Böse*. Naumann.

Nyholm, S. (2020). Humans and robots: Ethics, agency, and anthropomorphism. Rowman and Littlefield.

OpenAI. (2022). Dall-E 2. https://openai.com/dall-e-2/ (Last Accessed June 15, 2022).

Ryle, G. (1949). *The concept of mind*. CUP.

Sartre, J.-P. (1993). *Being and nothingness*. Washington Square.

Smith, J. K. (2021). *Robotic persons: Our future with social robots*. WestBow.

Vallor, S. (2012). Flourishing on facebook: Virtue friendship & new social media. *Ethics & Information Technology, 14*, 185–199. https://doi.org/10.1007/s10676-010-9262-2

Vallor, S. (2018). *Technology and the virtues*. OUP.

CHAPTER 10

Almost a Conclusion

This investigation set out under the premise that an investigation into human–machine friendships is worth the effort. I believe we have been proven right. Not only has been made clear that there are good arguments against such friendships and even better ones in favor, but we have also seen that the exchange of such arguments will help us better understand the nature of friendship. In this way, just as proposed in the first chapter of this investigation, the debate around human–machine friendships is related to the one surrounding robot rights. In both discourses reign harsh opinions that appear to represent wider philosophical beliefs and theories. Both may trigger strong emotional responses on the basis of these beliefs and theories, for the spectrum of positions goes to the extremes. And yet, in both discourses, the main takeaway from all debates is that we learn most about the concepts once we apply them to machines. As Gunkel alluded to, the debate surrounding robot rights provides a surface to reflect on our concept of ‚rights' rather than merely just an answer to the question whether we should apply rights to robots (cf. Gunkel, 2018).

This is the case for friendship, too. It is not a surprise that the philosophically most common approach to friendship is 2500 years old, invented in a highly misogynist, patriarchal, and hierarchical society from which Aristotle was not free himself. Since then, adding people to the circle of those who can be virtue friends has been extended, from poor

© The Author(s), under exclusive license to Springer Nature   203
Switzerland AG 2022
H. Kempt, *Synthetic Friends*,
https://doi.org/10.1007/978-3-031-13631-3_10

men, to enslaved men, to women. Our understanding of what a friendship can be has been expanded time and time again, yet the normative order, the idea that philosophers could determine what a virtue friendship or "the most excellent friendship" looks like, remains. It remains as a guiding hypothesis to test whether we could or should create machines to be virtue friends with us. John Danaher, Sven Nyholm, Alexis Elder, Margaret Archer, and others take it as a given to attempt to incorporate machines into this normative framework without questioning whether the framework finally reaches its frame-limit. It was invented as a tool to govern, sanctifying the social networks of privileged men, and since then has been slowly adapted to include more and more individuals. Maybe it was never the correct tool for analyzing human friendships to begin with.

We took the emergence of human–machine friendships as an opportunity to reflect on the concept of friendship. We found that we fare best if we provide negative conditions for friendships and merely positive suggestions for friendships, subtly leaving the attempt to force human–machine friendships into this frame behind. We might want to consider virtue friendships something inherently human (Kempt, 2021), something first and foremost concerned with reconstructing how our friends make us better people. This does not necessarily transfer to the friendships possible with machines, and thus an application to them might fail.

We have seen that attaching ourselves to mainstream ideas of what an excellent friendship makes renders the attempts to transfer them to machine relationships riddled with concerns, thus not helping our philosophical goal, to provide a comprehensive understanding of how to be friends with a machine, grow.

It even might fail to transfer to our new technologically mediated lifestyles, though many people still may disagree. The normativity of immediateness, the imperative to make do with the people around you, is being replaced with the idea of normativity of opportunity: with the ability to connect with almost anyone on earth at our fingertips, we should question what has been put on us before. Digital hermits, as a lifestyle for many, is also reflective of these processes.

The negative conditions for human–machine friends mostly concern the temptations of technology and counter the negative consequences of power imbalances and our vulnerabilities in intimate relationships. The positive suggestions encourage us to accept that we get to create: what has been deemed narcissistic, arrogant, or delusional, as we should not want to create "better" entities. With the negative conditions in place and some sincerity about the wish to have help at being better, and using technology to improve our lives already, we considered that using the unique features of social machines can massively help our lives. From keeping old memories accessible to keeping our secrets without judgment, from expressing sympathy to staying around for as long as we need them, machines can do what even the best human friends cannot.

Ultimately, we found that a distinctive proposal of the philosophical possibility of human–machine friendships, accounting for the differences in kind of humans and machines, should be reflected in the name we give certain social machines: synthetic friends. Synthetic friends are exactly those social machines that observe the negative conditions of friendship and yet are capable of being socially integrated in someone's life.

The biggest concern in this context, following these comforting, quality-of-life-improving features, consisted in us preferring the company of those machines over real human beings. This, so went the argument, could eventually lead to us fully abandoning human connections, or lose the ability to tolerate human idiosyncrasies in favor of the unlimited patience of a machine. We argued that friendships are fundamentally socially integrated, and thus, human–machine friendships should as well. In this sense, then, social integration might be the biggest hill to climb for human–machine friendships. Not only because we encounter a diverse set of attitudes, potentially conflicting cultural narratives and traditions about the role of technology in private life, and political agenda-setting that could easily target early adopters of human–machine friendships but also because of the specific social norms we unconsciously expect from other human friends.

We proposed to not only create machines that can fulfill these norms but also that some of those norms should be changed or even abandoned for machine friends. Not because otherwise technology will never reach those levels, but rather because otherwise, we will push people to hide their human–machine friendships. Such a proposal is not intended to be a blueprint, but an elaboration on the conditions that must be met for human–machine friendships to grow. Yet, formulating conditions should provide us with the good conscience that once those conditions are met, the relationships between humans and machines we encounter are truly friendships in a philosophically demanding, non-exclusionary sense.

What is left are criticisms of this approach. We discussed the concern about watering down the concept of friendship by lowering the necessary condition for it while leaving the sufficient condition untouched.

Yet, we pointed out that these concerns really just apply to the academic exercise of finding the correct name for a phenomenon, and synthetic friends might cover this concern. For the actual lived experience, people will not care for the specific name, and will probably come up with their own words for them. At the same time, many people have no concern in calling people friends they barely know to communicate their familiarity with them or even pets.

That there are some unspoken issues, either from "vibes" or from existential worries, has been acknowledged but is more representative of specific friendships others may not ever have. It would thus be an odd requirement for machines to be able to fully realize these rather rare features of friendships. Also, our approach is laying the groundwork for expanding the family-resemblance-project of friendship to include machine–machine friendships and others without having to declare these events absurd.

Further, we can understand their fundamental difference to be a chance for expanding our minds and social expectations without atrophying or corroding our abilities to relate to other humans.

10.1   Onward?

Some kind of human–machine relationships will happen that people will insist to be friendships. Philosophy may best be served in aiming to explain these lived realities. This is not helped by an insistence on realism that condemns these realities as shallow, delusional, incomplete, or simply wrong.

This is, ultimately, a conservative judgment of denial. Just because one would have created the future differently does not equip one to demand a different present. A philosopher's job, in this regard, is not only to demand a better world in the future but also to make sense of the present. Reconstructing the range that the concept of friendship can take, and enabling people to lead friendships with machines, is to fulfill this very task: not only to help those building the future to create a better one but also for those who live the present to understand it.

Maybe controversially, but certainly as a result of philosophical reasoning, we have concluded that human–machine friendship can emerge: reliably, deeply, and satisfyingly, while expanding our horizons, with synthetic friends.

REFERENCES

Gunkel, D. (2018). *Robot rights*. MITP.

Kempt, H. (2021). *Zwischenmenschlichkeit für Maschinen*. In KI–Die Große Verheißung. Xenomoi.

Index